03WSM3

Protocols for Studying Wet Weather Impacts and Urbanization Patterns

by:

Christine A. Pomeroy
University of Utah (formerly with Colorado State University)

Larry A. Roesner
Department of Civil and Environmental Engineering
Colorado State University

James C. Coleman II
Camp Dresser & McKee Inc.

Ed Rankin
Midwest Biodiversity Institute

2008

WERF

The Water Environment Research Foundation, a not-for-profit organization, funds and manages water quality research for its subscribers through a diverse public-private partnership between municipal utilities, corporations, academia, industry, and the federal government. WERF subscribers include municipal and regional water and wastewater utilities, industrial corporations, environmental engineering firms, and others that share a commitment to cost-effective water quality solutions. WERF is dedicated to advancing science and technology addressing water quality issues as they impact water resources, the atmosphere, the lands, and quality of life.

For more information, contact:
Water Environment Research Foundation
635 Slaters Lane, Suite 300
Alexandria, VA 22314-1177
Tel: (703) 684-2470
Fax: (703) 299-0742
www.werf.org
werf@werf.org

This report was co-published by the following organizations. For non-subscriber sales information, contact:

IWA Publishing
Alliance House, 12 Caxton Street
London SW1H 0QS, United Kingdom
Tel: +44 (0) 20 7654 5500
Fax: +44 (0) 20 7654 5555
www.iwapublishing.com
publications@iwap.co.uk

Library of Congress Catalog Card Number: 2007923910

IWAP ISBN: 978-1-84339-787-8

Colorado State University

The research on which this report is based was developed, in part, by the United States Environmental Protection Agency (EPA) through Cooperative Agreement No. CR-831559-01 with the Water Environment Research Foundation (WERF). However, the views expressed in this document are solely those of Colorado State University and neither EPA nor WERF endorses any products or commercial services mentioned in this publication. This report is a publication of WERF, not EPA. Funds awarded under the Cooperative Agreement cited above were not used for editorial services, reproduction, printing, or distribution.

ACKNOWLEDGMENTS

The project team would like to acknowledge numerous individuals who provided assistance in various areas of the project.

Cathy Tate, Elise Giddings, Jeff Steuer, and Thomas Cuffney
United States Geological Survey

Donna Neubert and Mitch Heineman, P.E.
CDM

City of Raleigh, North Carolina

Town of Cary, North Carolina

Report Preparation:

Principal Investigators:
Larry A. Roesner, Ph.D., P.E., *Colorado State University*
Christine A. Pomeroy, Ph.D., P.E., *University of Utah (formerly with Colorado State University)*
Brian P. Bledsoe, Ph.D., P.E., *Colorado State University*

Project Team:
James C. Coleman II, Ph.D., *Camp Dresser & McKee Inc.*
Ed Rankin, *Midwest Biodiversity Institute*
Boris Kondratieff, Ph.D., *Colorado State University*
Jennifer Davis, *Colorado State University*
John Edgerly, *Colorado State University*
Ivan Rivas, *Colorado State University*

Project Subcommittee

Peter Ruffier, *City of Eugene, Ore., PSC Chair*
Ben Urbonas, P.E., *Urban Drainage and Flood Control District, Co-PSC Chair*
Faith Fitzpatrick, Ph.D., *United States Geological Survey*
Nancy White, MLA, Ph.D., *UNC Coastal Studies Institute*
Phil Kaufmann, Ph.D., *United States Environmental Protection Agency*

Water Environmental Research Foundation Staff:

Director of Research: Daniel M. Woltering, Ph.D.
Program Director: Margaret Stewart
Program Manager: Jane M. Casteline

ABSTRACT AND BENEFITS

Abstract:

This research explores the relationship of urbanization to ecology in the wadeable streams of developing watersheds and describes a protocol, which urban stormwater management agencies can apply in their jurisdiction, to prevent the severe ecologic deterioration that presently occurs in urban streams as a watershed develops. The protocol comprises an eight-step process that involves data collection and analysis integrated with mathematical modeling of runoff from existing and proposed urban development to determine how biologic health indices, such as the benthic index of biotic integrity (B-IBI) or Ephemeroptera, Plecoptera, Trichoptera (EPT) richness, respond to urbanization of a watershed. Stormwater managers can apply this protocol in their areas of jurisdiction to assist them in developing urban runoff management rules and design criteria with some assurance that the resulting development will minimize negative impacts on the ecology of receiving streams. In addition to the full protocol, a simple protocol is also provided for municipal agencies to use that does not require extensive data collection.

Benefits:

- Sets criteria for designing runoff controls on new development with some assurance that the resulting development will significantly reduce the hydrologic, geomorphic, and ecologic impacts on receiving streams;
- Refines current drainage criteria for flood control and stormwater management, thereby eliminating the need to make drastic changes to current drainage system design criteria; and
- Provides guidance for data collection in support of the protocol to ensure the proper number and type of physical, chemical, and biologic data are included in the sampling program and that the timing and spatial scale of the sampling of all types of data are compatible.

Keywords: Urban stormwater management, aquatic ecology, water quality, flood control, BMPs.

TABLE OF CONTENTS

LIST OF TABLES

LIST OF FIGURES

EXECUTIVE SUMMARY

ES.1 Introduction and Background for This Study

Land use change, especially urbanization, has profound effects on the runoff characteristics of land that it occupies and, consequently, on the aquatic environments of the urban streams to which that runoff drains (WEF & ASCE, 1998). In a previous study for WERF, Roesner and Bledsoe (2003) explored the state of knowledge with respect to the effects of wet weather flows on the physical character of aquatic habitat and identified knowledge gaps with respect to causal relationships. They then examined the comprehensiveness of existing data and made a qualitative determination about the usefulness of those data for further analysis to increase knowledge of the effects of wet weather flows on the physical character of aquatic habitat. Finally, they recommended future research studies that would increase knowledge in the subject area, with emphasis on pilot-scale projects that could be used to develop practical protocols for preventing or mitigating detrimental effects of urban development on receiving streams. The scope of their study was limited to the investigation of the impacts of wet weather flows from *separate* urban storm drainage systems on physical habitat of streams in urbanized watersheds.

Based on the findings and recommendations of that study, WERF funded a research study to develop a protocol that municipalities responsible for stormwater management could use to evaluate the impacts of land use patterns and alternative stormwater management strategies on the biotic integrity of streams in urbanizing watersheds. The study called for development of a protocol followed by an application of the protocol to urbanization questions in a pilot study area.

ES.2 Basis for the Protocol

The protocol is based on two key findings:

- A key finding from the literature review is the demonstration by Booth et al. (2004) that in the Pacific Northwest, the biotic integrity of an urban stream as measured by the benthic index of biological integrity (B-IBI) can be related directly to the hydrologic metrics, $T_{0.5}$ and T_{Qmean}, and *that these two metrics are in turn related to the intensity of urbanization on a watershed.* The $T_{0.5}$ hydrologic metric was defined as the fraction of the time, over a multiyear record, during which the flow in the stream equals or exceeds the peak flow of the 0.5-year storm generated by existing conditions in each watershed. The T_{Qmean} was the fraction of time, over a multiyear record, that the flow is greater than the mean or average flow.
- A key research finding from this project by Edgerly et al. (2006) is that the hydrologic metric $T_{0.5}$ can be computed using mathematical stormwater models, *and that* $T_{0.5}$ *is sensitive to alternative land use and runoff control scenarios.*

This means that if mathematical models can be used to generate hydrologic metrics, such as $T_{0.5}$ and T_{Qmean} for alternative land use patterns and/or runoff control strategies, and information that relates fish or macroinvertebrate metrics to hydrologic metrics can be developed for a region of interest, then land use planning and runoff control strategies can be related to biotic integrity in streams. This relationship is illustrated pictorially in Figure ES-1.

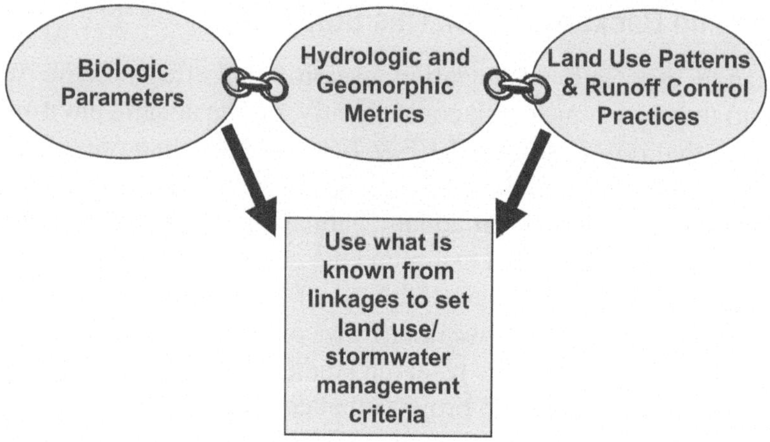

Figure ES-1. Linking Land Use Patterns and Runoff Control Practices to Stream Biologic Integrity Through Hydrologic and Geomorphic Metrics.

ES.3 Overview of the Protocol

Building on these two key findings, the protocol illustrated in Figure ES-2 was developed. The black boxes portray the identification and collection of necessary biologic data;

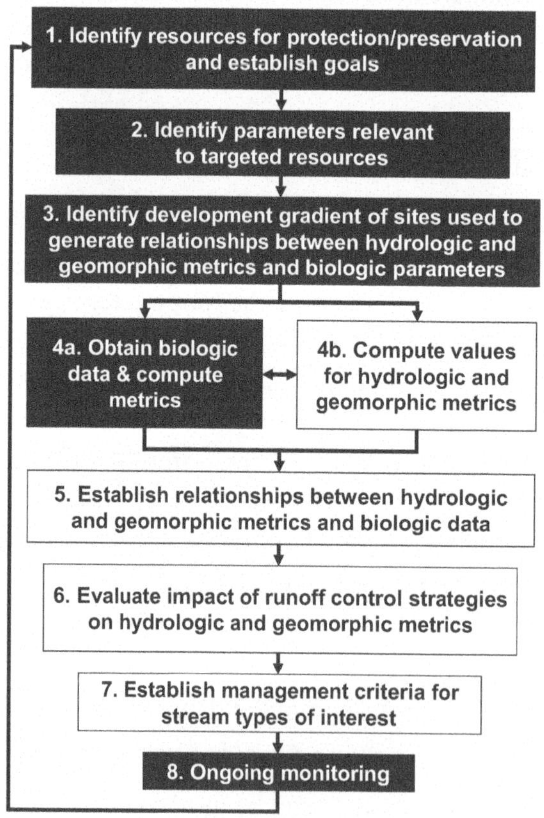

Figure ES-2. The Protocol.

detailed guidance on these steps is contained in Chapter 5.0. The white boxes in Figure ES-2 are data analysis activities comprising hydrologic, geomorphic, and biotic analyses that define the baseline for stormwater management planning and form a basis for evaluating the relative impact of alternative stormwater management plans (land use patterns and/or runoff control strategies) on stream biota for developing watersheds. Details of these analytical tasks are contained in Chapter 3.0 of this report. It should be noted that the focus of the protocol is on the relationship of stormwater management practices in urbanizing watersheds to biologic health in the receiving streams, as represented by measures of the fish and aquatic macroinvertebrate communities. Water quality is not explicitly included in the protocol; however, *it is included implicitly* in the protocol because the protocol requires stormwater treatment practices be employed as part of the hydrologic stability analysis.

The eight steps of the protocol are summarized below. Details can be found in the main body of the report.

Step 1 involves the setting of protection goals for the target streams, which may be performed in the context of existing state or tribal water quality standards.

Step 2 constitutes the identification of the biological, chemical, and/or hydrologic and geomorphic parameters to be measured.

Step 3 involves identification of study sites representing the full range of development conditions (from undeveloped to urban) in support of establishing stressor–response gradients for the study area.

Step 4a involves the collection of biologic data for each study site.

Step 4b is the first of the analytical procedures in the protocol and requires continuous flow records (≥ 5 years) for the computation of hydrologic and geomorphic metrics. The continuous flow records can be obtained from long-term monitoring or through hydrologic and hydraulic modeling.

Steps 5 and 6 utilizes the data collected in Steps 1 through 4 of the protocol to develop linkages among hydrologic and geomorphic, biologic metrics, and the land use/level of development in study watersheds.

Step 7 uses these linkages in the analysis and development of stormwater management criteria for the protection of stream resources.

Step 8 provides guidance for ongoing monitoring to ensure that established management criteria are sufficiently protective of the target water resources. If monitoring studies indicate degradation in stream conditions relative to the acceptable management goals, an adaptive management approach is required. In this case the study protocol may be repeated or more refined indicators may be selected (e.g., Barbour et al., in press) to refine linkages among the development gradient and type and water resource condition. Refined linkages could then be used to revise and implement management criteria to be followed during additional ongoing monitoring.

ES.4 Testing of the Protocol in the North Carolina Piedmont

The protocol was tested in the North Carolina Piedmont (Raleigh-Durham metropolitan area) using data gathered by the U.S. Geologic Survey (USGS) under the National Water Quality Assessment (NAWQA) program study *Effects of Urbanization on Stream Ecosystems* (EUSE)

(Couch and Hamilton, 2002). In all, 30 watersheds were examined for appropriateness, and eight were selected to complete the analysis. The Raleigh, North Carolina, region was chosen for application and testing of the protocol based on the size of the study watersheds and the availability of data, which were collected from the summer of 2002 to the fall of 2003. Nearly half of the 30 watersheds evaluated in the North Carolina study are less than 5 square miles, a size that is practical for modeling at the level of detail needed to produce data useful for calculating hydrologic and geomorphic metrics. This watershed size is also the typical size that a stormwater management agency would look at for application of the protocol.

ES.4.1 Step 1 — Identify Resources for Protection/Preservation and Establish Goals

As noted above, this protocol is a planning tool for stormwater managers to use in urbanizing watersheds to help them protect and/or limit the degradation of the existing stream biota as land use in the watershed changes from rural to urban. The protocol is a tool for examining the effects of alternative land use patterns and urban runoff control measures on stream hydrology, geomorphology, and aquatic biota. Because this application is a test of the protocol rather than a planning application, this step was not performed; rather it was assumed that the resources for protection would be the existing stream hydrologic and geomorphic condition in an undeveloped watershed.

ES.4.2 Step 2 — Identify Parameters Relevant to Resources to Be Protected

Based on the literature review and research from this study, it was determined that benthic macroinvertebrate metrics would be the best indicators of instream biotic integrity in the North Carolina Piedmont. Under the USGS EUSE study, macroinvertebrates were collected using two types of samples — a quantitative sample collected from instream riffles (richest targeted habitat, RTH) and a qualitative multihabitat (QMH) sample that collected invertebrates from as many habitats in the stream reach as were accessible (Giddings & Steuer, 2007). In Step 2 these various indices were examined for the most appropriate index to use for the protocol application to the test area.

ES.4.3 Step 3 — Identify Development Gradient of Sites Used to Generate Relationships between Hydrologic and Geomorphic Metrics and Biological/Habitat Parameters.

Thirty North Carolina watersheds were screened against the site selection criteria developed in this project including: homogeneous land use, streams with similar geomorphic characteristics, watershed size, and no confounding factors such as point source discharges or dams. Watersheds were also selected based on the degree of urbanization in each basin as represented by multimetric urban intensity index (UII) values computed by the USGS. The UII takes into account census, socioeconomic, infrastructure, land-use, and land-cover characteristics correlate with population density. From the 30 candidate watershed sites, eight were selected for application of the protocol; these watersheds range in size from 3 to 9 square miles. Table ES.1 lists the selected NAWQA study watersheds, their corresponding drainage area size, and their associated level of urbanization/development as represented by the UII.

ES.4.4a Step 4a — Obtain Biological Data for Each Site and Compute Biologic Metrics

Normally, the watersheds selected in Step 3 would be the watersheds on which sampling of the macroinvertebrate of fish parameters identified in Step 2 would occur. In this study, the data collected by the USGS in the EUSE study (USGS, 2007) were used. EPT richness values were computed from the USGS EUSE macroinvertebrate data. The B-IBI values were computed using methods described by Kerans (1994) for rivers of the Tennessee Valley. A B-IBI method is not available specifically for the North Carolina Piedmont ecoregion.

Table ES.1. Selection Characteristics of Watersheds Chosen for Protocol Application.

Watershed	USGS Site Number	Basin Size (sq mi.)	Urban Index (UII)	Total Impervious Area (%)	Homo-geneity (0=Low 1=Med 2=High)	Streambed Material % Silt / Clay / Sand	Streambed Material % Gravel / Cobble
Strouds Creek	0208501535	8.9	7.2	3.8	1	3.6	81.9
Morgan Creek	02097464	8.3	4.9	0.8	2	16.3	58.2
Camp Branch	0208794025	3.2	12.9	2.9	1	16.4	78.1
Dutchmans Branch	0208758440	4.6	28.6	4.4	1	9.1	90.9
Wilson Creek	0209750881	3.5	24.1	9.0	1	34.6	63.6
Beaverdam Creek	02087304	3.0	90.8	11.7	2	40.0	58.2
Pigeon House Creek	0208732610	4.5	100.0	30.6	2	30.9	58.2
Black Creek	0208725055	3.6	79.6	20.5	2	23.6	16.3

ES.4.4b Step 4b — Compute Values of Hydrologic and Geomorphic Metrics for Each Site

Based on the findings of Booth et al. (2004) in the Pacific Northwest, and the research involving preliminary testing of the protocol as part of this study with respect to hydrologic metrics in Fort Collins, Colorado and Atlanta, Georgia (Edgerly et al., 2006), the hydrologic $T_{0.5}$ metric was selected as the hydrologic and geomorphic metric measure of urbanization. *In the analyses presented in this study, the $T_{0.5}$ is defined as the percent of time that the flow is greater than the peak flow of the 0.5-year storm,* rather than the fraction of time that the flow is greater than the peak flow of the 0.5-year storm used by Booth et al. (2004). The 0.5-year storm is defined from a partial duration series of flows.

It should be noted that other metrics may be more appropriate for use in other regions of the country. If possible, a number of hydrologic metrics should be evaluated to identify those that most strongly correlate to indicators of biotic integrity. Climatic region and stream type should be taken into consideration when determining which hydrologic metrics to evaluate. Olden and Poff (2003) presented a list of recommended indices that best represent the biologically critical portions of the flow regime, and these are provided in the Appendix of this report.

Figure ES-3 shows that the $T_{0.5}$ hydrologic metric, as expected, is strongly correlated to the UII for the eight watersheds selected for application of the protocol. The reason the range of values for this metric is different for the record *measured* by the USGS during the EUSE study (ranging from 12–17 months) compared to the *simulated* flow record (20-year record), is that the simulated flow record contains the response of a larger range of storms than the measured flow record. Simulated stream flow values were generated using the U.S. Environmental Protection Agency's (EPA's) Storm Water Management Model, SWMM5, which was calibrated to each of the eight watersheds using the 12–17 month measured records, then used to simulate long-term stream flows using 20 years of 15-minute rainfall data from a rain gauge in the pilot study area.

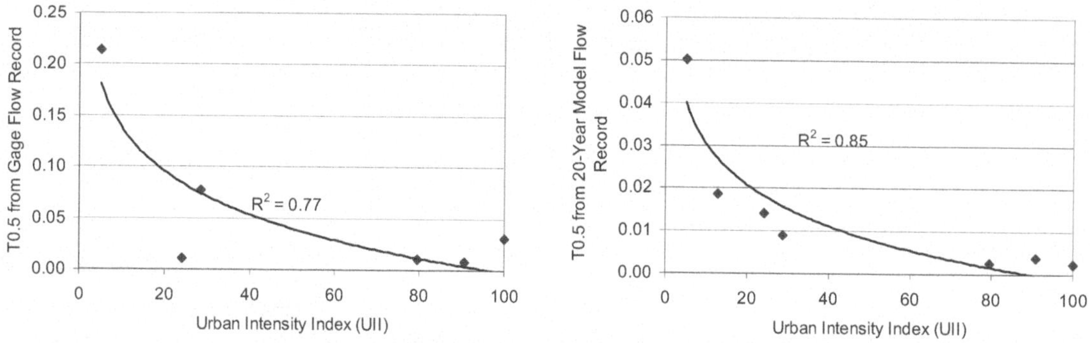

Figure ES-3. Relationship of UII to $T_{0.5}$ Metrics Calculated from flows from the USGS Stream Gauges in the Raleigh Durham, North Carolina Area and from SWMM Modeling.

Geomorphic metrics describing cumulative excess shear stress and erosion potential were calculated using water depth output from the SWMM models at 15-minute time intervals. The geomorphic metrics were computed for the predominant bed substrate size class identified for each watershed during the USGS EUSE studies. Predominant categories ranged from sand to large cobble.

ES.4.5 Step 5 — Establish Relationships between Hydrologic and Geomorphic Metrics and Biologic Data

This step integrates the biologic data collected by the USGS EUSE study in Step 4a with the hydrologic metric data calculated in Step 4b. Figure ES-4 shows regressions of the $T_{0.5}$ to the four macroinvertebrate indices selected as indicators of stream health (see Step 2 for definition): Ephemeroptera, Plecoptera, and Trichoptera (EPT) richness (RTH); EPT richness (QMH); EPT *percent* richness (RHT); EPT *percent* richness (QMH); and Benthic Index of Biotic Integrity (B-IBI). The $T_{0.5}$ metrics from a number of different time periods (1-, 2-, 5-, 10- and 20-years) were computed from a 20-year record of stream flows simulated with the SWMM5 runoff model. The $T_{0.5}$ hydrologic metric values computed from the 2- and 10-year flow records were the best predictors of macroinvertebrate response. In general, the strongest relationships between biologic results and $T_{0.5}$ values were observed for the B-IBI and for EPT richness for both the RTH and QMH. The $T_{0.5}$ hydrologic metric was consistently a strong predictor of ETP richness for the QMH across all time periods examined, indicating that it is the best overall macroinvertebrate metric to relate to the $T_{0.5}$ hydrologic metric in the North Carolina Piedmont region. The strong relationship between B-IBI and the $T_{0.5}$ hydrologic metric is similar to results found by Booth et al. (2004).

Relationships between cumulative excess shear stress and macroinvertebrate metrics were evaluated, but it was determined that bed substrate was a better predictor of macroinvertebrate response. Because substrate was independently a better predictor of macroinvertebrate response than cumulative excess shear, the predominant substrate should be used as a geomorphic metric to predict macroinvertebrate response rather than cumulative excess shear. The computation of cumulative excess shear, however, is necessary for determining erosion potential, which is a useful tool in evaluating stormwater management scenarios.

ES.4.6 Step 6 — Evaluate Impact of Runoff Control Strategies on Hydrologic and Geomorphic Metrics.

The Morgan Creek watershed (see Table ES-1) is the most undeveloped of the watersheds examined in this study. The impervious area is essentially the paved road network through the watershed. This watershed was selected to demonstrate that through the use of hydrologic and hydraulic models (analogous to SWMM5) it is possible to estimate the $T_{0.5}$ for changes in land use and/or application of stormwater management practices, and hence the impact on stream biota (as measured by the selected macroinvertebrate metrics). The land use of the Morgan Creek watershed was changed to the same level of imperviousness as the Pigeon

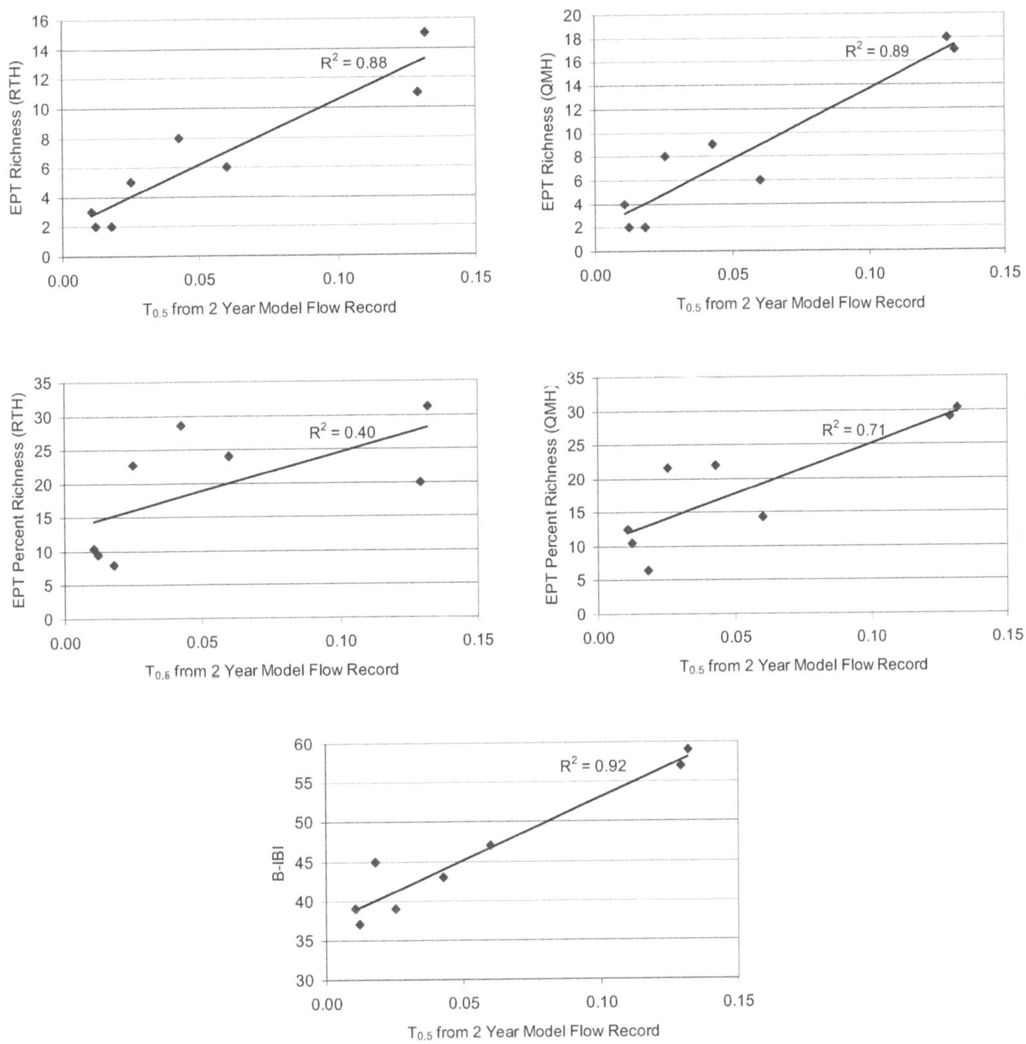

Figure ES-4. Relationship of $T_{0.5}$ Calculated From the 20-Year Flow Record to Benthic Macroinvertebrate Sampling Data (a) EPT Richness (RTH), (b) EPT Richness (QMH), (c) EPT Percent Richness (RTH), (d) EPT Percent Richness (QMH), (e) Benthic Index of Biotic Integrity (B-IBI).

House Creek watershed, the most highly urbanized of the watersheds studied (see Table ES-1). Initially, 20 years of runoff with the changed land use were simulated with SWMM5 assuming no runoff controls because Pigeon House Creek watershed is the old central city area of Raleigh and has no urban runoff controls. The $T_{0.5}$ was then computed for the Morgan Creek watershed under the changed land use and compared to the $T_{0.5}$ for Pigeon House Creek watershed.

It was discovered that changing land use in the Morgan Creek watershed did not change the value of the $T_{0.5}$ hydrologic metric as much as was expected. The metric value fell in the range of those observed for partially developed watersheds (Wilson Creek) or those developed at low-to-medium intensity (Dutchman's Branch), rather than near the metric value of Pigeon House Creek watershed, as was expected. When the watershed channel cross-section geometry was changed to reflect the observed channel shape in watersheds with the same land use as simulated in Morgan, the $T_{0.5}$ changed in the expected manner. The reason is easily seen in Figure ES-5; the rural cross-section is typical of the shape presently found in the undeveloped Morgan Creek watershed, while the urbanized cross-section is taken from the Pigeon House Creek watershed with an upstream drainage area of the same size as the cross-section shown for the Morgan Creek watershed. Given the significant differences in channel geometry, it is expected that the $T_{0.5}$ is affected by the channel geometry.

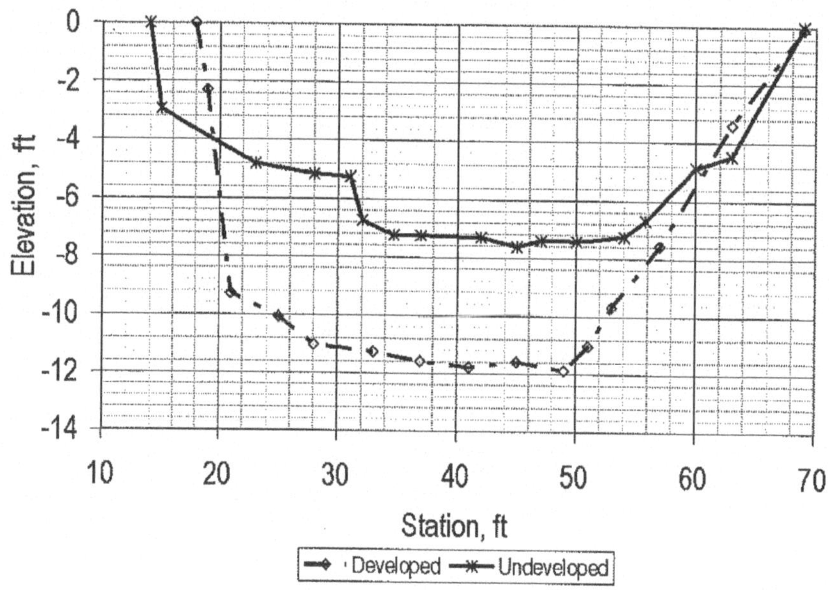

Figure ES-5. Representative Cross-Section for Rural (Undeveloped) and Urbanized (Developed) Channel Geometry in the Morgan Creek Watershed.

In addition to developing protocols for studying wet weather impacts and urbanization patterns, an objective of this study was to develop methodologies to identify stormwater management control measures that minimize degradation to the biotic integrity of streams. To examine this question, the land use of the Morgan Creek watershed was changed to the same land use intensity in the Pigeon House Creek watershed, but runoff practices were applied to to each catchment within the watershed at the catchment outfall.

The runoff management practices applied to the changed land use condition on Morgan were:

- Detention of the runoff from the 10-year storm with the changed land use to control the peak release rate to the peak flow from the 10-year storm with the current land use and channel geometries.

- Detention of the runoff from the 2-year storm with the changed land use to control the peak release rate to the peak flow from the 2-year storm with the current land use and channel geometries.
- An extended detention BMP sized according to the WEF/ASCE Manual of Practice on Urban Stormwater Quality Management (WEF and ASCE, 1998), with a 40-hour drawdown. The extended detention BMP was designed to capture of 70-90% of all runoff-producing events in their entirety; this design is intended to provide water quality treatment of runoff.

The location of these runoff control features is shown in Figure ES-6. For this runoff control scenario, SWMM5 was used to simulate a record of 20 years of 15-minute stream flow data from which the T0.5 was computed at the mouth of the watershed.

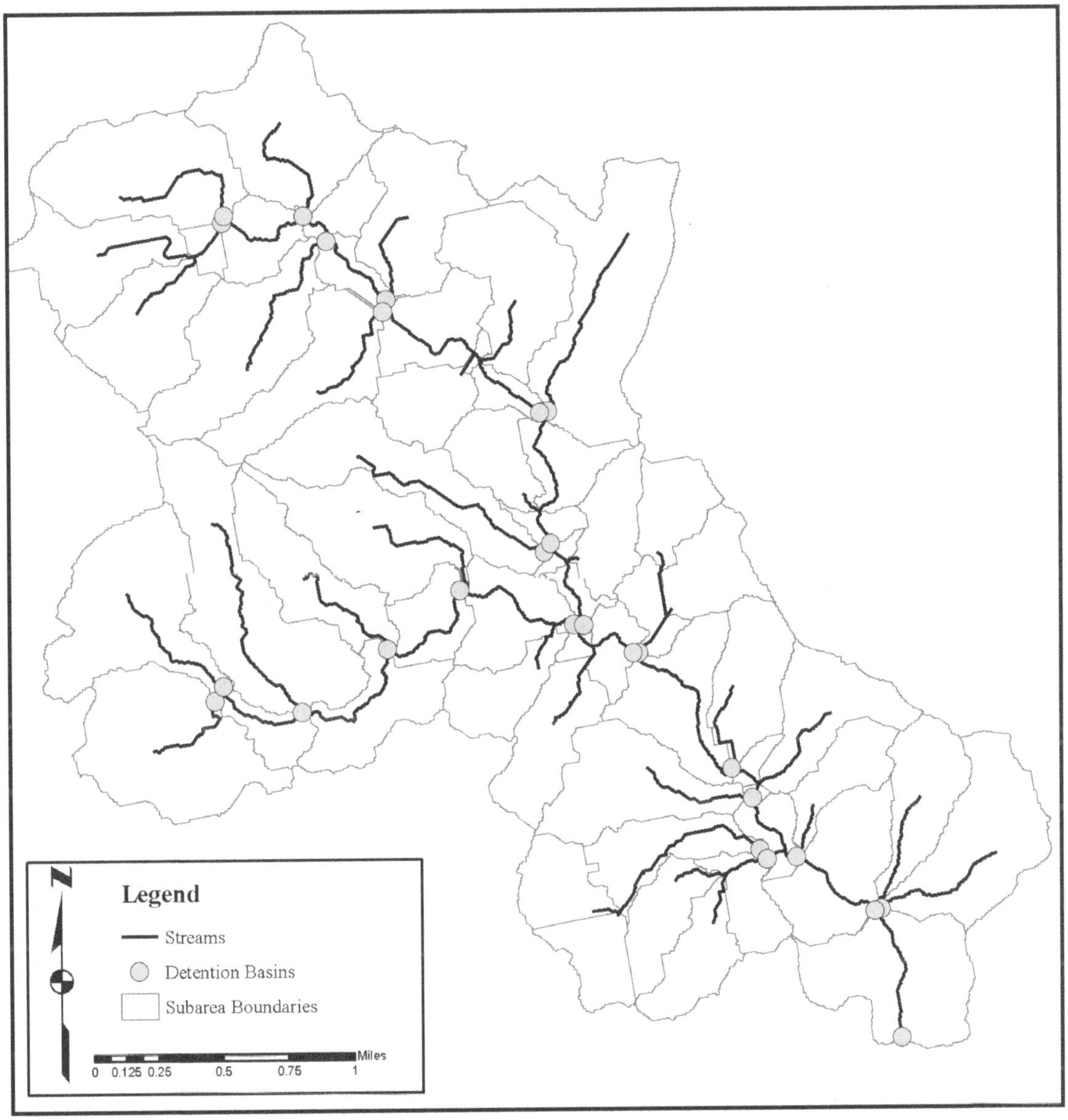

Figure ES-6. Location of Detention Basins Modeled in Morgan Watershed.

In addition to showing existing conditions, developed conditions without runoff controls are identified for scenarios where the existing channel geometry was held constant, and where the developed watershed channel geometry was applied. Lastly, the impact of stormwater controls on the $T_{0.5}$ hydrologic metric was evaluated in the Morgan Creek watershed for developed land use and current (undeveloped) channel geometry. Figure ES-7 shows that if development occurs in the Morgan Creek watershed without stormwater controls, but channel geometry remains constant, the EPT richness score will be expected to degrade to a fair rating from it's original good-fair rating. It is important to note, however, that changes in channel

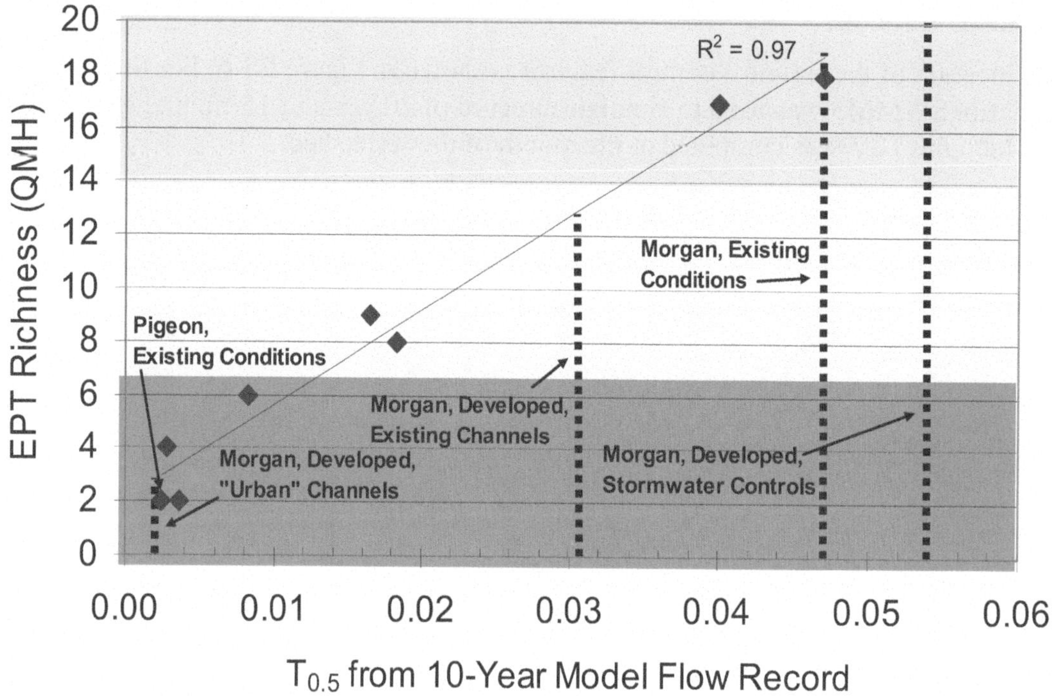

Figure ES-7. Change in the $T_{0.5}$ Metric with Changed Land Use and Stormwater Management Practices.

geometry are likely to occur if stormwater is left uncontrolled, given that increases in erosion potential are expected to occur (see below). In the case of the developed watershed with developed channel geometry it can be seen that the resulting $T_{0.5}$ will be nearly the same as that of Pigeon House Creek, and we can expect that the resulting biotic integrity of Morgan Creek will be similar to that which is currently observed in Pigeon House Creek, i.e. the lowest EPT richness score of the eight watersheds examined in this study. However, if the runoff controls described above are implemented as development occurs, we can expect the $T_{0.5}$ to change very little, and thus the biotic integrity of Morgan Creek should remain relatively unchanged, which is the desired endpoint of the protocol.

Table ES-2 shows the effect of stormwater management practices on erosion potential using cumulative excess shear stress values to the 1.5 power for one of the predominant size classes in the Morgan Creek watershed, very coarse gravel (d_s>32 mm). The erosion potential analysis was conducted for rural channels, only, as it is not appropriate to relate cumulative excess shear stress values computed for channels with different geometry. Table ES-2 shows that a desired erosion potential value of 1.0 is achieved when erosion potential is computed for the 20-year period of record; a slight over control of erosion potential occurs for the 10-year periods of record. It should be noted that the erosion potential computations do not account for changes in upland sediment supply that occur with land use change in a watershed or the application of stormwater management BMPs. Stream channel stability is dependant on a balance between sediment transport capacity of a stream and the supply of sediment both from upland and in-stream sources. Therefore, erosion potential values of less than 1.0 do not necessarily mean that the flows in the channel are overcontrolled.

Table ES-2. Summary of Erosion Potential for the Morgan Creek Watershed.

Scenario	10-Year	20-Year
Developed, No Controls	1.24	1.54
Developed, Stormwater Controls	0.93	0.99

ES.4.7 Step 7 – Establish Management Criteria for Stream Types of Interest

After completing the types of analysis discussed in Step 6, it will be possible to set stormwater management criteria for development on an urbanizing watershed that reduces the negative impact of urbanization on the biota in the receiving stream. Step 7 is the administrative procedure of codifying the stormwater management criteria into the existing drainage criteria manuals.

Note that this study did not look at the impact of stormwater runoff from construction sites, but rather focused on post-development runoff. If construction site runoff is not managed well, especially with respect to Total Suspended Solids in the discharges, the post-development condition of the stream will be severely impacted.

ES.4.8 Step 8 – Ongoing Monitoring

This task might be called the adaptive management task of the protocol because it allows for the adjustment of criteria set based on application of the protocol if they are not providing the desired biotic outcome. Ongoing monitoring should take place during and after construction on representative watersheds to ensure that the biotic integrity in the stream remains at the expected health level. If deterioration beyond what is expected occurs, then the analysis steps that took place in the original application of the protocol should be revisited to determine if any of the assumptions used might be changed as a result of more recent experience. If so, the effect of those modified assumptions on stormwater management criteria should be evaluated and the drainage criteria changed as appropriate. Although this adaptive management process may be too late for protection of the existing watershed, as monitoring will likely show degradation after development has taken place, the adaptive management process may be used to identify retrofit opportunities or can be used to minimize negative impacts of urbanization due to development of other watersheds in the future.

ES.5 Limitations of the Protocol

It is important for the reader to understand that the protocol presented in this report is not the last word in evaluating the impacts of urban runoff on the biotic integrity of streams, but rather it is a tool or methodology that can be used to help stormwater managers and planners to better understand the potential impacts of implementing alternative structural runoff control strategies on stream in urbanizing watersheds. It does not *predict or simulate* the biotic health of streams, but rather provides an estimate for the relative degradation in the biotic integrity of streams that might be expected if certain runoff control strategies are implemented. The strength and the weakness of this protocol is that its implicit assumption is that the effect of urban runoff structural controls on in-stream biotic integrity is entirely hydrologic. Geomorphology and water quality are not considered, except as they might be affected by the changed hydrology due to runoff controls or pollutant removal that occurs in BMPs used to control the runoff rate from small storms.

Of additional concern regarding the protocol is the use of SWMM and other hydrologic and hydraulic models which have the ability to evaluate the hydrologic impacts of land use change and stormwater management practices. These models should be used with caution for the computation of hydrologic and geomorphic metrics of biologic interest. Many factors influencing historical or natural flow regimes (i.e., large woody debris, beavers, wetlands) as well as factors influencing current and future flow (i.e., influence of groundwater withdrawals on summer baseflows) may not be accurately estimated by these models (Cassin et al., 2005).

The power of the protocol is that it focuses on minimizing the change in the hydrologic regime of the watershed runoff in the urbanized state from that of the pre-development state. Moreover, the full hydrologic spectrum of runoff is considered, from very small storms that transport most of the pollutant load off the watershed to the 100-year storm; this is done by considering how both the peak-flow frequency curve and the flow-duration curve are modified by urbanization and how different runoff control strategies change those two curves compared to the pre-development curves. What we have learned from previous studies is that by using the full spectrum hydrologic control approach, it is possible to develop runoff control strategies in urbanizing watersheds that reproduce the pre-development peak flow frequency curve, and most of the pre-development flow-duration curve, using state of practice peak shaving (detention) controls and a volumetric BMP. We have discovered further that this “full spectrum control” advocated in this report also significantly reduces the erosive work that occurs in the stream channels of urbanized watersheds. Intuitively, if the hydrologic regime of the runoff from an urbanized area is controlled so that it reproduces the pre-development runoff regime of the watershed, then the geomorphic and biotic stream degradation will be minimized.

There are some issues that are not addressed by the protocol that the user should be aware of; these are:

- The protocol deals only with post-urbanization conditions, it does not address the effects of changes in sediment load from a watershed. It does not address issues related to increased sediment loads that might occur during development if the proper erosion controls are not applied.
- The protocol does not account for the fact that the watershed sediment load is usually less after urbanization than prior to urbanization due to changes in land use or the sediment trapping abilities of stormwater best management practices.
- The protocol does not take into account stream stabilization practices. These practices, alone or in conjunction with stormwater BMPs are an option to increase stream stability and the potential for biotic integrity since 100% control of runoff from any urbanizing area may not be physically and institutionally possible.
- There is a body of thought that stream benthic biota respond more to the geomorphology, i.e. changes in sediment size resulting from urbanization, than to the hydrology; but since changes in the geomorphology of the stream is a function of the hydrology of the stream, it may be that the use of the hydrologic stream metric is still a more suitable link between urbanization and the biotic integrity of streams.

- Water quality impacts on benthic macroinvertebrates are not addressed at all in the protocol. The implicit assumption is that since BMPs are used as part of the hydrologic control strategy, that if the proper BMP is used, the pollutants of concern will be removed. That assumption remains to be tested, however changes in hydrologic regime do play a role in geomorphic and biotic degradation of streams and control of changes in the hydrologic regime are critical. As better runoff control practices are implemented in urbanizing watersheds, the geomorphic and biotic conditions of the steams will improve, and it will then be time to turn attention toward the quality of the discharges in urban runoff.

The protocol is a step forward in developing tools that WERF Subscribers can use to for determining the effect of their runoff control practices in their jurisdictions on the biotic health of the streams in urbanizing watersheds. There is a large body of literature on the effects of urbanization on the geomorphology and the biotic integrity of receiving streams, but there are virtually no tools that stormwater managers can use to help them determine the best stormwater management practices to require of developers to minimize the effects of urbanization on the receiving streams. The protocol fills that knowledge gap, and, although not perfect, it is better than any other currently available tool. As additional knowledge is gained on the cause-effect relationships of flow, geomorphology, water quality, and biotic integrity in urban streams, these relationships can be incorporated into the model, improving its predictive capability and usefulness as a planning tool.

ES.6 Conclusions and Recommendations

The following conclusions are based on this project:

- Hydrologic metrics, which reflect the change in the hydrologic regime that occurs with urban development, have a higher correlation to the biotic integrity of a stream, as measured by macroinvertebrate indices, than indirect measures of urbanization such as percent imperviousness.
- Of the 9 hydrologic metrics examined by Edgerly et al. (2006) in preliminary testing of the protocol, the $T_{0.5}$ is the most responsive to changes in land use and mitigation through runoff controls.
- Physically based mathematical models that simulate runoff and stream flow (e.g., SWMM5) can be used to estimate the value of the hydrologic metric $T_{0.5}$ for alternative land use patterns and/or runoff control strategies in developing watersheds. This was done by first determining $T_{0.5}$ values for watersheds of varying levels of urbanization in the same region and comparing those $T_{0.5}$ values with those of urbanization watersheds. This allows stormwater managers to infer how alternative stormwater management plans and strategies will affect the biotic integrity of the receiving stream.
- The geomorphic impact of alternative stormwater management plans and/or runoff control strategies can be evaluated using computer simulations models (e.g., SWMM5) by using an instream erosion potential index.
- Research efforts to determine the biologic health of streams in study watersheds must reflect a data condition level of 3 or 4, generally associated with an intensive collection effort. A higher data condition level is required due to the inherent complexity of urbanization impacts on stream ecosystems. The study should optimally use multiple organism groups, such as fish and aquatic macroinvertebrates,

in order to provide a more robust approach for assessing urbanization impacts. Also, multiple individual metrics of the biological community would optimally be used, with each reflecting different environmental sensitivities. Utilizing multiple community metrics with differing sensitivities to environmental stressors provides a better reflection of the overall biotic integrity of the stream(s).

- Ecological monitoring data, for biological and habitat parameters, are essential to implementation of the protocol. The availability of data that meets data quality objectives for this type of work may not be readily available in all parts of the country, although state agencies and the U.S. EPA are working to refine monitoring programs and to develop tiered aquatic life uses. The choice of biological indicators and endpoints can have important consequences on study results, such as the example used here. Results from another recent WERF project (01-WSM-3, Barbour et al., in press) are particularly relevant for parts of the country in which adequate biological monitoring data are scarce or absent. The project's findings provided an approach for generating biologic indicators that are sensitive to the suite of urban stressors specific to a region or study area.
- In small communities that cannot afford to conduct the data sampling and computer modeling required by the protocol, the following minimum set of runoff control criteria for new development should be applied to provide improved protection of streams from a hydrologic, geomorphic, and biologic perspective:

    - The peak discharge from the 100-, 10-, and 2-year storm shall not exceed the peak discharges for these storms in the predevelopment condition.
    - An extended detention pond with at least a 40-h drawdown time that is sized according to the WEF/ASCE (1998) *Manual of Practice on Urban Stormwater Management* shall be required on new development. If local standards requiring longer draw-down times are in place, these should be implemented.

It should be borne in mind that the watersheds in the North Carolina Piedmont on which data were gathered do not have properly sized best management practices (BMPs) or runoff controls for small storms; nor do the watersheds reviewed in the literature as far as is known. This research assumed that the relationship of the macroinvertebrate indices to the $T_{0.5}$ is sensitive only to the absolute value of the $T_{0.5}$ and not to other hydrologic effects that may not be captured by the $T_{0.5}$, such as the protracted duration of flows smaller than the 0.5-year storm that result when extended detention storage with 40-h drawdown is applied to a watershed. It should be possible to indirectly assess the effect of protracted low flows using the instream erosion potential index because it is a measure of the geomorphic stability of the channel, and hence the integrity of the benthic community. But, this is engineering judgment, not proven fact. Therefore, it is recommended that a carefully conducted experiment be conducted on an urbanizing watershed of 1 to 3 square miles, with runoff controls and BMPs designed according to the criteria recommended in this study (see Conclusions above) to relate macroinvertebrate health to $T_{0.5}$ as the watershed develops and compare this data to watersheds that are in various states of urbanization but that have no significant number of runoff controls. Care must be taken in this experiment to control runoff during the construction phase or the macroinvertebrate indices will reflect the construction impacts rather than the built environment impacts.

It is believed that implementation of stormwater management practices that reduce the overall volume of runoff through infiltration or evapotranspiration will also aid in moving towards a more natural hydrologic flow regime that will allow for healthy receiving waters. The extent of the impact of these practices is dependent upon the extent to which they are implemented. These practices were not evaluated as part of this study, however, future research is needed to quantify the biotic impacts of volume reduction through practices such low impact development.

Water quality was not addressed in this work. From the standpoint of protecting receiving streams in urbanizing watersheds, the project team believes that water quality is not the key issue; the runoff controls required to achieve hydrologic and geomorphic stability must include BMPs which if properly designed will remove pollutants from stormwater runoff. Nevertheless, the federal Clean Water Act is based on water quality criteria in receiving streams and Stormwater Regulations now require that total maximum daily loads (TMDLs) include stormwater runoff. Therefore, it is recommended that future research build on this Protocol, adding water quality sampling of constituents identified in the water quality criteria applicable to the receiving stream and that stormwater models add water quality as a runoff parameter and simulate the removal of pollutants by runoff control practices. Available simulation models, including U.S. EPA's SWMM5 model, are a long way from being able to credibly simulate either runoff quality or constituent removal in BMPs.

CHAPTER 1.0

INTRODUCTION

1.1 Background for this Study

Land use change, especially urbanization, has profound impacts on the runoff characteristics of land that it occupies and, consequently, on the aquatic environments of the urban streams to which that runoff drains (WEF & ASCE, 1998). A review of the literature over the last 20 years reveals that research on the effects of urban runoff on aquatic biota has increased significantly during this period, but little effort has been made to relate the alteration of physical habitat in receiving waters to specific hydrologic and geomorphic changes on the landscape (i.e., land cover and drainage system hydraulic features). A typical example of the current approach to these kinds of studies is the study by Maxted and Shaver (1996), who studied the biotic integrity of stream systems downstream of urban best management practices (BMPs) at several sites in Maryland. Effects were related only to gross percent imperviousness of the tributary watershed and included little information on the outlet design criteria of the BMPs. Additionally, articles on the effects of urbanization on runoff have tended to concentrate on the hydrology and hydraulics with only categorical references to receiving water impacts. See for example Roesner and Burgess (1991) and Roesner (1997), who present detailed analyses of the hydrologic effects of BMPs on urban discharge flow frequency curves in various cities but do not include corollary data on biotic impacts.

The fundamental hydrologic changes that tend to be associated in varying degrees with urbanization include the following: more frequent and higher magnitude flows (Bledsoe and Watson, 2001; Booth and Jackson, 1997; Leopold, 1968); increased duration of geomorphically significant flows (Hollis, 1975; MacRae, 1997; McCuen & Moglen, 1988); flashier/less predictable flows (Henshaw & Booth, 2000; Konrad, 2000); altered timing and rate of change relative to riparian and floodplain connections (Poff et al., 1996), altered duration of low-flow periods (Finkenbine et al., 2000; Konrad, 2000), and conversion of subsurface distributed discharge inputs to surface (point) discharge (Booth & Henshaw, 2001). However, all of these observed phenomena are interrelated (Roesner & Bledsoe, 2003).

In a previous study for the Water Environment Research Foundation (WERF), Roesner and Bledsoe (2003) explored the state of knowledge with respect to the effects of wet weather flows on the physical character of aquatic habitat and identified knowledge gaps with respect to the ability to define the cause–effect relationships. They then examined the comprehensiveness of the data used in support of the published literature in the subject area and made a qualitative determination of the usefulness of those data for further analysis to increase knowledge in the subject area. Finally, they recommended future research studies that would increase knowledge in the subject area with emphasis on pilot-scale projects that could be used to develop practical protocols for preventing or mitigating detrimental effects of urban development on receiving streams. The scope of the study was limited to the investigation of the impacts of wet weather flows from *separate* urban storm drainage systems on physical habitat in urban streams within an urbanized watershed.

A comprehensive search of the literature and data pertaining to physical habitat in urban streams was undertaken in that study. The literature search targeted articles from appropriate peer-reviewed journals and easily accessible gray literature such as conference proceedings published by the Water Environment Federation (WEF), WERF, the American Society of Civil Engineers (ASCE), the International Water Association (IWA), and the U.S. Environmental Protection Agency (U.S. EPA). Personal communication was also used to identify additional references on wet weather flow impacts. More that 550 articles were reviewed. More than 180 "useable" literature and data sources, namely information that could be used to further the understanding of the relationships between urban wet weather flows and physical habitat, were tabulated with respect to the scope of the problem that was covered by the individual publications. This effort revealed, however, that very few monitoring programs and research initiatives are *quantifying* physical habitat in a manner that permits direct analysis of the multiscale process linkages. Notable exceptions are the U.S. EPA Environmental Monitoring and Assessment Program (EMAP) (Kaufmann et al., 1999) and the National Water Quality Assessment (NAWQA) Program within the United States Geological Survey (USGS) (Fitzpatrick et al., 1998; Jacobson et al., 2001; Fitzpatrick et al., 2005).

1.1.1 Major Findings and Conclusions of the Previous Study

The literature search and personal inquiries produced the following observations:

- Local descriptions of stream substrates, flow conditions, and pool characteristics, combined with watershed-scale indices of human disturbance, such as imperviousness and population, are the standard explanatory variables in assessing stream biotic integrity and habitat quality. The linkages between larger-scale processes and the small-scale habitat descriptors are not well documented and poorly understood.
- The current framework for interpreting biomonitoring data tends to operate at two disparate spatial scales; i.e., ecoregions and snapshots of local habitat. For example, in the U.S. EPA *Biological Criteria: Technical Guidance for Streams and Small Rivers* (Gibson et al., 1996), flow regime is identified as a key factor at *all spatial scales,* yet it is typically described only locally.

1. Not only is the flow measured at the local scale, but it is just a snapshot of the actual flow regime taken at the time of sampling.
2. Stream habitat simplification as a result of modified wet weather flow regimes is a reality, but recognition of this phenomenon is mostly based on qualitative and anecdotal evidence.

Certain patterns in the literature were readily seen. First, of all the papers reviewed, less than 10% truly integrated hydrology, geomorphology, and multiscale physical habitat. Of those, about 30% were literature reviews and not actual studies involving data gathering. Many of the research publications covered two categories well at one spatial scale (e.g., reach-scale, geomorphology and physical habitat) but did not touch upon the other categories. Some of the research addressed urbanization and physical habitat but did not address the hydrologic/geomorphologic connections between. Finally, there was relatively little attention given to system level response and consequent effects on quantity and quality of microhabitat. As a result of these findings, the following key gaps in the literature were identified:

1. We lack a solid conceptual framework for predicting the impact of large-scale watershed modifications and wet weather flows on processes that influence stream communities;
2. A need exists for longer-term monitoring;
3. There is no widely accepted system for quantifying geomorphic instability and degradation of physical habitat;
4. There is a need for *process-based* stream classification;
5. Specific links between urbanization characteristics and stream degradation are lacking;
6. There is a need for urban best management practice (BMP) assessment standards; and
7. Developing a multiscale understanding of *habitat potential* in human-dominated watersheds is needed.

1.1.2 Recommendations from the Previous Study

The recommended research program included first and foremost comprehensive, long-term monitoring augmented with mathematical modeling of the linkages among development style/drainage system design, flow regime, and multiscale changes in physical habitat and biotic response. Improved diagnosis and predictive understanding of future change requires multifaceted, multiscale, and multidisciplinary studies based on a firm understanding of the history and processes operating in a drainage basin (Jacobson et al., 2001). Detailed long-term analyses of the influence of hydrologic regime and channel morphology on differences between communities in recruitment, immigration/emigration, mortality, and age structure are also needed. Finally, future research should directly examine tradeoffs involving:(1) flood mitigation versus channel roughness, habitat heterogeneity, debris inputs, and riparian protection; 2) chemical water quality improvement through extended detention versus geomorphically based flow regime controls; and 3) rehabilitation of aquatic habitat using static features versus allowing potential dynamic adjustments in channel form and habitat structure.

1.2 The Basis for This Study

Based on the findings and recommendations of the study described in the previous section, WERF decided to fund a research study to develop a protocol that municipalities responsible for stormwater management could use to evaluate the impacts of land use patterns and alternative stormwater management strategies on the biotic integrity of streams in urbanizing watersheds. The study called for the development of the protocols followed by a test application of the protocols to a real watershed.

Building on the extensive literature review from the previous study, the literature review for this study focuses on advancements in research regarding the linkages between urbanization, hydrology, hydraulics, geomorphology, physical habitat, and the biotic integrity of streams that have occurred since 2002. Older materials are included only when they are critical to the explanation of current research or if they were not included in the previous review. This review covers three major areas of research that are important to the establishment of these linkages: 1) developments in refining urbanization-aquatic ecosystem linkages, 2) mechanisms of altered flow regimes, and 3) ecosystem flow and habitat requirements. Key findings from this literature review and a research project conducted as part of this project formed the basis of the protocol:

- The key finding from the literature review is that Booth et al. (2004) demonstrated that the biotic health of an urban stream as measured by the benthic index of biological integrity (B-IBI) in the Pacific Northwest is directly related to the hydrologic metrics, $T_{0.5}$ and T_{Qmean}, and *that these two metrics are in turn related to the intensity of urbanization on a watershed.* In this protocol, the $T_{0.5}$ hydrologic metric is defined as the percentage of the time, over a multiyear record, that the flow in a stream equals or exceeds the peak flow of the 2-yr storm, and T_{Qmean} is the percentage of time, over a multiyear record, that the flow is greater than the mean or average flow.

- A key piece of research from this project (Edgerly et. al., 2006) found that the hydrologic metric $T_{0.5}$ can be computed using mathematical storm drainage models, *and that* $T_{0.5}$ *is sensitive to alternative runoff control scenarios.*

If mathematical models can be used to generate hydrologic metrics such as $T_{0.5}$ and T_{Qmean} for alternative land use patterns and/or runoff control strategies, and if information can be developed in planning areas that relates biologic metrics to hydrologic metrics, then land use planning and runoff control strategies can be related to the biotic integrity in streams. This relationship is illustrated pictorially in Figure 1-1.

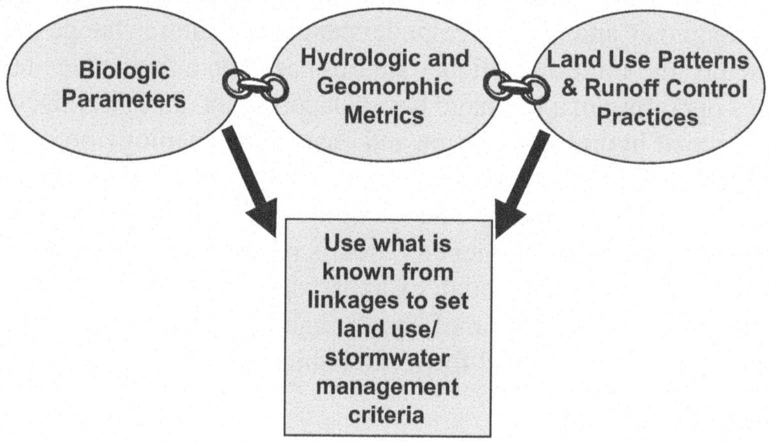

Figure 1-1. Linking Land Use Patterns and Runoff Control Practices to Stream Biologic Integrity through Hydrologic and Geomorphic Metrics.

1.3 Overview of the Protocol

Building on the findings described above, the protocol illustrated in Figure 1-2 was developed. The black boxes comprise the identification and collection of necessary biologic and habitat data; detailed guidance on these steps is contained in Chapter 5.0. The white boxes in Figure 1-2 are data analysis activities comprising hydrologic, geomorphic and biologic analyses that define the baseline for stormwater management planning and form a basis for evaluating the relative biotic impact of alternative stormwater management plans (land use patterns and/or runoff control strategies) for developing watersheds. Details of these analytical tasks are contained in Chapter 3.0 of this report. It should be noted that the focus of the protocol is on the relationship of stormwater management practices in urbanizing watersheds to biological health in the receiving streams as represented by measures of the fish and aquatic macroinvertebrate communities, and physical habitat condition. Water quality is not explicitly included in the protocol; however, *it is included implicitly* in the protocol because the protocol requires stormwater treatment practices be employed as part of the hydrologic stability analysis.

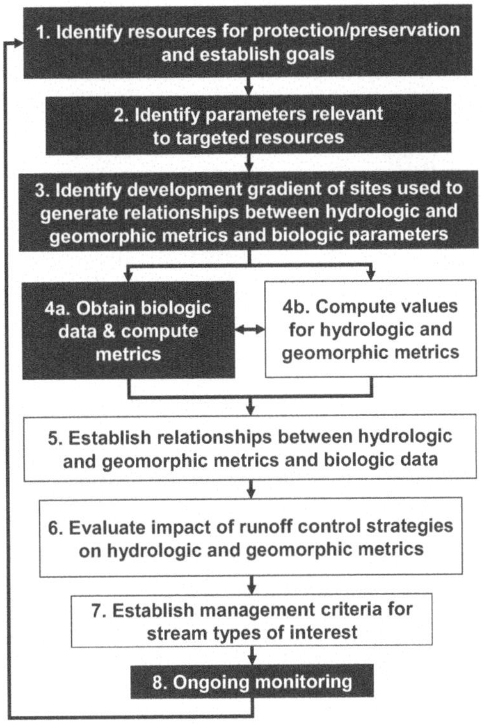

Figure 1-2. Overview of the Protocol.

The eight steps of the protocol are summarized below. Details can be found in subsequent sections of the report.

Step 1 involves setting of protection goals for the target streams, discussed in greater detail in a following section, which may be performed in the context of existing state or tribal water quality standards.

Step 2 is the identification of the biological, chemical, and/or hydrologic and geomorphic parameters to be measured.

Step 3 involves the identification of study sites representing the full range of development conditions (from undeveloped to urban) to establish stressor-response gradients for the study area.

Step 4a involves the collection of biological and habitat data for each study site.

Step 4b requires continuous flow records for the computation of hydrologic and geomorphic (stream) metrics. The continuous flow records can be obtained from long-term monitoring or through hydrologic and hydraulic modeling.

Steps 5 and 6 utilize the data collected in Steps 1 through 4 of the study protocol to develop linkages among hydrologic and geomorphic, biological, and habitat metrics and the land use/level of development in the selected study watersheds.

Step 7 uses these linkages in the analysis and development of stormwater management criteria for the protection of stream resources.

Step 8 provides guidance for ongoing monitoring to ensure that established management criteria are sufficiently protective of the target water resources. If monitoring studies indicate degradation in stream conditions relative to the acceptable management goals, an adaptive management approach is required. In this case the study protocol may be repeated or more refined indicators may be selected (e.g., Barbour et al., draft 2006) to improve linkages among the development gradient and type and water resource conditions. Refined linkages could then be used to revise and implement management criteria to be followed during additional ongoing monitoring.

1.4 Limitations of the Protocol

It is important for the reader to understand that the protocol presented in this report is not the last word in evaluating the impacts of urban runoff on the biotic integrity of streams, but rather it is a tool or methodology that can be used to help stormwater managers and planners to better understand the potential impacts of implementing alternative structural runoff control strategies on streams in urbanizing watersheds. It does not *predict or simulate* the biotic health of streams but rather provides an estimate for the relative degradation in the biotic integrity of streams that might be expected if certain runoff control strategies are implemented. The strength and the weakness of this protocol is its implicit assumption that the effect of urban runoff structural controls on in-stream biotic integrity is entirely hydrologic. Geomorphology and water quality are not considered, except as they might be affected by the changed hydrology due to runoff controls, or pollutant removal that occurs in BMPs used to control the runoff rate from small storms.

The power of the protocol is it focuses on minimizing the change in the hydrologic regime of the watershed runoff in the urbanized state from that of the predevelopment state. Moreover, the full hydrologic spectrum of runoff is considered, from very small storms that transport most of the pollutant load off the watershed to the 100-year storm, by considering how both the peak-flow frequency curve and the flow-duration curve are modified by urbanization and how different runoff control strategies change those two curves compared to the predevelopment curves.

What has been learned from previous studies is that by using the full spectrum hydrologic control approach, it is possible to develop runoff control strategies in urbanizing watersheds that reproduce the pre-development peak flow frequency curve, and most of the predevelopment flow-duration curve, using state of practice peak shaving (detention) controls and a volumetric BMP. It's been discovered further that this "full spectrum control" advocated in this report also significantly reduces the erosive work that occurs in the stream channels of urbanized watersheds. Intuitively, if the hydrologic regime of the runoff from an urbanized area is controlled so that it reproduces the predevelopment runoff regime of the watershed, then the geomorphic and biotic stream degradation will be minimized.

There are some issues that are not addressed by the protocol that the user should be aware of; these are:

- The protocol deals only with post-urbanization conditions, it does not address the effects of changes in sediment load from a watershed. It does not address issues related to increased sediment loads that might occur during development if the proper erosion controls are not applied. The protocol does not account for the fact that the watershed sediment load is usually less after urbanization, than prior to urbanization due to changes in land use or the sediment trapping abilities of stormwater best management practices.

- The protocol does not take into account stream stabilization practices. These practices, alone or in conjunction with stormwater BMPs, are an option to increase stream stability and enhance the potential for biotic integrity because 100% control of runoff from any urbanizing area might not be physically and institutionally possible.
- There is a body of thought that stream benthic biota respond more to the geomorphology (i.e., changes in sediment size resulting from urbanization) than to the hydrology, but because changes in the geomorphology of the stream are a function of the hydrology of the stream, it could be that the use of the hydrologic stream metric is still a more suitable link between urbanization and the biotic integrity of streams.
- Water quality impacts on benthic macroinvertebrates are not addressed at all in the protocol. The implicit assumption is that because BMPs are used as part of the hydrologic control strategy, that if the proper BMP is used, the pollutants of concern will be removed. This assumption remains to be tested; however, changes in hydrologic regime do play a role in geomorphic and biotic degradation of streams, and the control of changes in the hydrologic regime is critical. As better runoff control practices are implemented in urbanizing watersheds, the geomorphic and biotic conditions of the steams will improve, and it will then be time to turn attention toward the quality of the discharges in urban runoff.

The protocol is a step forward in developing tools that WERF subscribers can use for determining the effect of their runoff control practices in their jurisdictions on the biotic health of streams in urbanizing watersheds. A large body of literature exists on the effects of urbanization on the geomorphology and the biotic integrity of receiving streams, but there are virtually no tools that stormwater managers can use to help them determine the optimal stormwater BMPs to require of developers to minimize the effects of urbanization on receiving streams. The protocol fills that knowledge gap, and, although it is not perfect, it is better than any other currently available tool. As additional knowledge is gained on the cause-effect relationships of flow, geomorphology, water quality, and biotic integrity in urban streams, these relationships can be incorporated into the model, improving its predictive capability and usefulness as a planning tool.

1.5 Testing the Protocol

The protocol was tested in the North Carolina Piedmont (Raleigh-Durham metropolitan area) using data gathered by the USGS under the NAWQA program study *Effects of Urbanization on Stream Ecosystems* (USGS, 2007). In all, 30 watersheds were examined for appropriateness and eight were selected for testing.

1.6 Organization of This Report

The centerpiece of the protocol is the analytic approach used to link land use patterns and/or runoff control patterns to biotic response. Therefore, Chapter 3.0 addresses the protocol for linking urbanization to water environment parameters (i.e., hydrology, geomorphology, and biologic metrics). Chapter 2.0 provides an overview of recent literature that describes linkages between urbanization and environmental parameters. Chapter 4.0 describes the application of the protocol in the North Carolina Piedmont and shows the relationship of hydrologic and geomorphic metrics to land use changes and runoff controls. Finally, biological and habitat sampling protocols are provided in Chapter 5.0.

CHAPTER 2.0

RECENT (2002 TO 2005) LITERATURE REVIEW OF LINKAGES AMONG URBANIZATION, FLOW REGIME, HABITAT, AND BIOLOGICAL RESPONSE

2.1 Introduction

The intent of this literature review is to build upon the extensive (more than 550 articles) review conducted for the WERF report by Roesner and Bledsoe (2003), *Physical Effects of Wet Weather Flows on Physical Habitats: Present Knowledge and Research Needs.* This was done by focusing on recent advancements in research regarding the linkages between urbanization, hydrology, hydraulics, geomorphology, physical habitat, and stream ecology that have occurred since 2002. Older materials are included only when they are critical to the explanation of current research or if they were not included in the previous review. The review covers three major areas of research that are important to the establishment of these linkages: 1) ecosystem flow and habitat requirements, 2) developments in refining urbanization-aquatic ecosystem linkages, and 3) mechanisms of altered flow regimes.

2.2 Ecosystem Flow and Habitat Requirements

Maddock et al. (2004) discussed the complex interaction of four primary factors in determining habitat in aquatic systems: water quality, the energy budget, the physical structure of the channel, and the flow regime. The combination of the latter two factors results in the instream physical habitat for biota in lotic ecosystems. In addition to these four factors, sediment sources and dynamics, combined with the physical structure of the channel and flow regime, determine habitat. Sediment dynamics are determined in part by hydraulics, but are also determined by a combination of watershed and local geologic settings, local channel boundary conditions, and stream network position (Fitzpatrick & Peppler, 2007). With regard to flow regime, a major trend in research on determining the optimal flows is approaching the problem from the bottom-up, one species at a time. Studies on the flow preferences of individual species have been conducted for decades (Statzner & Moss, 2004).

Shear stress has also recently gained attention in the ecohydraulics community (Tharme, 2003). A hydrologic change that increased shear stress or reduced the thickness of the viscous sublayer would reduce the availability of suitable microhabitats for some species, while increasing habitat availability for other species. This alters community functioning and food-web dynamics (Gore et al., 2001). Merigoux and Doledec (2004) also examined the relationship between near-bed hydraulics and benthic invertebrate response. With direct measurements of near-bed hydraulic conditions and samples of invertebrate taxa distribution, they were able to show relationships between hydraulic parameters and taxon richness, biological traits, and functional diversity using simple linear regressions. Shear stress and Froude number were found to be the most important hydraulic parameters, while substrate particle size and bed roughness appeared to be less influential (Merigoux & Doledec, 2004).

In addition to considerations of the hydraulic component of physical stream habitat, there is a substantial body of literature that provides anecdotal but important links between aquatic species and specific physical habitat features and recently more quantitative information linking physical features to the presence and abundance of various aquatic taxa (e.g., Kaufmann et al., 1999). Maddock (1999) pointed out that key research issues in physical stream habitat evaluation lie in decisions regarding the appropriate level of detail to provide a meaningful, yet cost-effective investigation, as well as identifying which factors are truly biologically relevant rather than simply geomorphic features. The U.S. EPA's Environmental Monitoring and Assessment Program (EMAP) has been a driving force in obtaining much quantitative habitat information and recently has performed studies in both the western and eastern halves of the United States. When planning for the EMAP physical data collection effort, U.S. EPA met with experts in these areas and developed a list of seven general categories of physical habitat attributes thought to be important influences on the biota of streams (Kaufmann, 1993). These included:

- Stream size — channel dimensions;
- Channel gradient;
- Channel substrate size and type;
- Habitat complexity and cover;
- Riparian vegetation cover and structure;
- Anthropogenic alterations; and
- Channel-riparian interaction.

These are similar to the categories used in Ohio EPA's qualitative habitat evaluation index (QHEI) (Rankin, 1989; 1995). U.S. EPA EMAP efforts have contributed to the recent focus on obtaining data linking these physical features to community level features in a quantitative manner. Often either the biological or habitat data have been relatively weak or not quantified in a repeatable fashion. Data collected over the last 10 years are beginning to result in quantitative data that should add precision to identify gradients of biological response to various physical habitat features.

Recent habitat modeling methodologies are commonly based on the integration of hydrologic, hydraulic, and biological response data such as model depth, velocity, substratum, cover, and benthic shear stress. These data are used to develop habitat-discharge curves for biota, which are then used to predict optimum flows (Tharme, 2003). Current modeling approaches addressing stream/river flow regimes and their effects on habitat and associated biota have their basis in fishery restoration efforts in flow-regulated rivers. Probably the most widely used method for analyzing linkages between flow and physical habitat for the purpose of formulating management options is the instream flow incremental methodology (IFIM), designed by the U.S. Fish and Wildlife Service (USFWS) for river system management (Bovee et al., 1998) and the associated Physical Habitat Simulation (PHABSIM) software (Milhous et al., 1984). PHABSIM is an extensive set of programs designed to predict the potential microhabitat conditions in rivers as a function of discharge and the relative suitability of those conditions to aquatic life based on empirical preference curves (Maddock, 1999). An underlying assumption of the methodology is that fish population dynamics are directly related to habitat availability and that habitat availability is directly related to a river's flow regime. It is debatable as to whether biota use habitat in proportion to its availability because biotic interactions may be the ultimate determinant of habitat use (Jacobson et al., 2001).

In a review of applications of the IFIM and PHABSIM over the past 20 years, Gore et al. (2001) identified five hydraulic conditions that most affect the distribution and ecological success of lotic biota: suspended load, bedload movement, water column effects (e.g., turbulence), velocity profile, and near-bed hydraulics. Their review noted that differences between predicted optimal and minimum flow criteria suggest that basing a management decision on a few target invertebrate species may result in the loss of habitat for other nontarget species. A larger proportion of benthic species can be provided habitat by using flows that are appropriate for functional feeding group targets or for highest community diversity. If the targets of high community diversity among benthic macroinvertebrates or the appropriate proportion of functional feeding groups are met, it will likely ensure a diversity of hydraulic conditions within the stream reach of interest. Producing suitability criteria for highest levels of benthic community diversity is a relatively simple procedure (Gore et al., 2001).

PHABSIM has also been used to examine linkages between flow and numerous variables affecting aquatic habitat and biota. Wu and Wang (2002) examined the relationship between substrate variation and flow in the framework of physical habitat simulation. The results imply that flow-related habitat studies are not usually sensitive to substrate alteration. Osmundson et al. (2002) demonstrated how flow regime can affect multiple aspects of an aquatic ecosystem through an indirect chain of linkages. They established statistical relationships between extended duration of fine sediment accumulation due to loss of flushing flow by a dam, depressed macrophyte production, reduced macroinvertebrate community integrity, and the health and survival of the pike minnow. Maddock et al. (2004) investigated the impact of channel morphology and flow on physical habitat availability for the two-spined blackfish (*Gadopsis bispinosus*) using PHABSIM. They found that the configuration of the channel bed topography is influential for the hydraulic habitat present at a given discharge and the response of this hydraulic habitat to discharge.

Building upon an established understanding of hydraulic habitat requirements of target fish species, Maddock et al. (2004) determined that streams with varying channel bed topography provide wide varieties of usable hydraulic habitat for the target species under the range of modeled flows. Excessively high water velocities were identified as the hydraulic habitat feature limiting success of the target species; therefore, research and modeling efforts were able to provide insight to physical stream habitat attributes limiting populations of the target species. Similarly, Booker and Dunbar (2004) used PHABSIM modeling to determine that greater suitable physical stream habitat was available over a wider range of hydraulic conditions in more natural river channels than in more engineered channels, thus contributing to the understanding of the interactions among physical channel characteristics, hydraulic conditions, and sustainable ecological communities in urban streams. Lamoroux et al. (1999), working in the Rhone River basin in France, sought to evaluate how hydraulic habitat structures fish communities. They concluded that hydraulics have a stronger effect on fish community structure within given geographic regions than if data for multiple geographic regions are considered together.

Thomson et al. (2004) also evaluated inter-regional differences in habitat-biota relationships when they examined the ecological significance of a geomorphic river characterization scheme by comparing macroinvertebrate assemblages and habitat characteristics of pool and run geomorphic units in three different types of rivers. The study tested the hypothesis that if geomorphic units are physically similar, they should also exhibit some biological similarities. Their results show significant correlation between variation in macroinvertebrate data and several local habitat variables across river types, suggesting that local factors are more important than regional factors.

Other researchers such as Lammert and Allan (1999) have evaluated habitat-biota relationships for hydraulic conditions along with other physical habitat characteristics at multiple spatial scales in an attempt to evaluate the relative importance of these factors in determining lotic communities. In a southeastern Michigan study, they observed that fish community composition showed a stronger relationship to flow variability and adjacent land use conditions, whereas macroinvertebrates were more strongly correlated with dominant substrate characteristics. More recently, Townsend et al. (2004) compared streams in pastureland and control streams in undeveloped, nonagricultural catchments with regard to macroinvertebrate community composition as well as to a number of physical channel characteristics, hydraulic variables, and physical habitat characteristics. Significant differences were observed between the two stream types with regard to water velocities, physical habitat conditions, and macroinvertebrate community composition and condition, though no conclusions were drawn regarding physical factors as determinants of the macroinvertebrate community in the study streams.

2.3 Developments in Refining Urbanization — Aquatic Ecosystem Linkages

It was noted in Roesner and Bledsoe (2003) that linking stream characteristics to gross measures of development, such as total imperviousness, percentage agriculture, or human population, provides little meaningful information for understanding key processes and creating practical mitigation strategies. In the past, studies have typically referred to the influences of urbanization in terms of percentage impervious area or percentage land cover, without specifically addressing land use types or stormwater management practices such as conveyance characteristics (e.g., pipes, swales), the use of detention, or BMPs. The dominant trend in these studies was to use simple linear regressions to link gross percentage of impervious area in a watershed with biological populations (Center for Watershed Protection, 2002; Roesner & Bledsoe, 2003). In order for the specific mechanisms, through which urbanization degrades aquatic ecosystems, to be identified and understood, accountability of hydrologic and geomorphic changes, land cover, and drainage system hydraulic features is needed (Roesner and Bledsoe, 2003).

A few recent studies have acknowledged the weakness of the common gross percentage impervious area as a predictor of aquatic community health and attempted to better characterize changes in land cover associated with urbanization in hope of finding more direct linkages to observed effects (Alberti et al., 2003; Booth et al., 2004; Walsh, 2004; Walsh et al., 2004). Booth et al. (2004) pointed out that the common definition of total impervious area (TIA) is hydrologically incomplete because it may overestimate pervious area in some cases by failing to account for compaction of soils and underestimate pervious area in other cases by failing to account for infiltration of runoff from small, isolated impervious surfaces that drain to adjacent pervious surfaces.

Total impervious area is not always an accurate predictor of effective impervious area. The degree to which impervious areas are connected to each other and ultimately to the stream is important to consider when characterizing urban development (Booth & Jackson, 1997; Alberti et al., 2003; Walsh, 2004). Direct measurement of directly connected impervious area (DCIA) is complicated (Booth & Jackson, 1997), however, and most studies do not differentiate between TIA and DCIA (Brabec et al., 2002). Lee and Heaney (2003) demonstrated the importance of distinguishing between TIA and DCIA when analyzing the hydrology of a system and showed that commonly used remote sensing techniques have limited accuracy for estimating DCIA. In the first part of their study, they used hourly long-term rainfall data and detailed estimates of land cover and hydraulic connection of surfaces in an apartment complex in Miami, Florida, to develop a linear rainfall-runoff estimation model. They found that DCIA made up 44% of the TIA but contributed 72% of the total runoff volume over 52 years. The second part of their study compared five levels of accuracy in the estimation of imperviousness for a development in Boulder, Colorado. The five imperviousness estimation methods with increasing levels of accuracy were as follows: 1) consulting literature reference data; 2) estimating TIA from a geographic information systems (GIS) database; 3) using a GIS database and subtracting isolated impervious areas from TIA to estimate DCIA; 4) using a GIS database and field investigation to include curbed streets in the DCIA estimate; 5) using a GIS database and field investigation for the entire area including rooftops with downspouts to estimate DCIA. The resulting estimates of DCIA obtained from these methods decreased significantly with increasing accuracy. As a result, runoff estimates generated by the EPA Stormwater Management Model (SWMM) 4.4 model using the values of imperviousness with the five levels of accuracy showed a significant decrease in the total volume of runoff predicted as accuracy of DCIA estimation increased. Lee and Heaney (2003) also found that transportation-related impervious area comprised 64% of the TIA and 97.2% of DCIA. This result is parallel with the findings of Alberti et al. (2003) who used multivariate analysis to show that road intensity is a good predictor of both the spatial arrangement of impervious surfaces and B-IBI scores.

Using GIS, Alberti et al. (2003) assessed the usefulness of urban pattern metrics for describing the spatial distributions, configuration, and hydraulic connectivity of landscape features (e.g., road density, number of road crossings per kilometer of stream, degree of fragmentation) as predictors of macroinvertebrate community health measured as the benthic index of biotic integrity (B-IBI). They found that road density and the number of road crossings are better predictors of B-IBI scores than total impervious area. They also found that the spatial arrangement of development patterns, particularly the amount and configuration of impervious surfaces and forest patches, plays an important role in determining the impact on stream macroinvertebrate communities and could explain much of the variability in B-IBI not explained by TIA. The results of their scale analysis suggest that basin-scale landscape configuration may influence the efficacy of local-scale riparian function.

Walsh (2004) measured macroinvertebrate assemblage composition across an urban density gradient and assessed whether the variance could be explained by percentage impervious area, drainage connection, septic tank density, catchment area, elevation, and geographic location. With respect to percentage imperviousness, typical patterns were observed: low catchment imperviousness was associated with diverse assemblages and high imperviousness was associated with disturbance-tolerant taxa. In the low-to-moderate range of imperviousness, however, there was a wide variation in assemblage composition and richness for which the strongest independent correlate was piped drainage connection. These results suggest that direct measurement of drainage connection is critical when estimating effective imperviousness. Walsh also recommended that efforts to restore streams in urban catchments should start with attention to the catchment drainage system. He suggests that conflicting reports among studies of dominant scale effects (local land use vs. catchment scale land use) could be explained by failure to account for differences in drainage connection.

In response to implications by researchers that riparian buffers were possibly the most important elements in protecting streams from the effects of urbanization (Horner et al., 2001), Walsh (2004) hypothesized that stormwater drainage system disconnect is of greater importance to aquatic biota than riparian buffers. In a study of the effects of urbanization on the geomorphology, habitat, hydrology, and fish index of biotic integrity of streams in the Chicago area, Illinois, and Wisconsin, Fitzpatrick et al. (2005) showed that the presence of stream buffers with forest and wetlands did not influence geomorphic, habitat, or hydrologic characteristics, or IBS scores. This suggests that forested riparian buffers are less able to moderate the influence of urbanization on biotic integrity of streams probably because of hydrologic alterations such as storm sewers, detention basins, and point-source discharges that directly bypass riparian buffers.

A number of researchers are now advocating low-impact development (LID) as an ecologically sound option for stormwater management, although most stress the fact that these new techniques do not eliminate the need for conventional, centralized stormwater management practices, such as detention (Konrad & Burges, 2001; May & Horner, 2002; Shaver & Ridley, 2002; Coffman, 2003; Holman-Dodds et al., 2003; Kirby, 2003). LID concepts for new development incorporate minimizing imperviousness, conserving native vegetation and soil, and implementing integrated, onsite treatment of stormwater (May & Horner, 2002). The theoretical advantage of such systems over a centralized stormwater management approach is that they manage the flow before it is concentrated in the drainage system, using spatially diffuse storage and release, a situation that is closer to the natural, pre-development runoff processes (Konrad & Burges, 2001).

One of the main design principles for "low-impact" new developments is clustered, compact development patterns, based on the assumption that compact urban forms, (smaller lot sizes and higher density) are more ecologically sound because they preserve open space (May & Horner, 2002; Booth et al., 2002; Alberti et al., 2003). There is little hard evidence, however, to support this assumption (Alberti et al., 2003; Center for Watershed Protection, 2003). Alberti et al. (2003) demonstrated a strong statistical correlation between increased percentage of impervious surfaces (and their aggregation) and ecological conditions in streams. In particular, as the probability of adjacency of paved areas rose from 50% to 100%, typical B-IBI values declined from 50 (excellent) to 10 (very poor). Although the findings clearly suggest that patterns of urban development affect watershed function, this relationship does not indicate a specific threshold. To further confound the issue, it was found that landscape configuration metrics (describing the degree of scatter and the proximity to the stream) are better predictors of B-IBI at the basin scale, whereas landscape composition metrics (i.e., percentage TIA and percentage trees) are better at the local scale. This observation supports the finding that scale is an important factor in assessing the effects of landscape patterns on ecological conditions.

2.4 Mechanisms of Altered Flow Regimes

After having discussed how urbanization changes the hydrologically significant aspects of land cover, a logical next step is to take a closer look at the characteristics of flow regimes, how flow regimes are altered by urbanization, and the ecological ramifications of such alterations. Hydrologic metrics that demonstrate altered streamflow regimes could provide a direct mechanistic link between aspects of urban development and degraded stream ecosystems (Booth et al., 2004).

Flow is a major driver behind many processes that ultimately determine the state of the stream ecosystem (Poff et al., 1997; Poff & Ward, 1989; Bunn & Arthington, 2002; Nilsson & Svedmark, 2002; Riis & Biggs, 2003; Roesner & Bledsoe, 2003); however, it can be very difficult to determine which mechanism is directly responsible for observed impacts in a stream (Bunn & Arthington, 2002). A major strength of management plans based on flow regime is that they have a low risk of overlooking flow requirements for important components of the ecosystem (Richter et al., 1996; Poff et al., 1997; Tharme et al., 2003; Suen et al., 2004). The five aspects of a flow regime identified as most important to the geomorphology, physical habitat, and ultimately stream biota are magnitude, frequency, duration, timing, and rate of change (Poff et al., 1997).

The recognition of the importance of flow regime to the ecological integrity of lotic systems has prompted several researchers to recommend measures for restoring a stream's hydrology to a condition more closely approximating the natural flow regime as part of overall ecological restoration efforts (e.g., Erskine, 1999; Ward et al., 2002; Poff et al., 1997; Richter et al. 1997, 1998; Richter & Richter, 2000; Richter et al., 2003). Though these recommendations are often directed toward restoration of natural flow regimes following dam construction and associated flow regulation, the principles underlying the recommendations are equally applicable to streams in urbanizing watersheds where flow regimes have been altered by development. Management efforts involving flow regime restoration are more tractable for regulated rivers, for which the allowed stream discharge may be more readily manipulated. For streams in urbanizing watersheds, however, the matter of flow regime stabilization or restoration is a more complex matter, involving runoff and flow mitigation in response to changes in land cover.

Probably the most widely used technique for quantifying changes in the flow regime caused by human activity is the indicators of hydrologic alteration (IHA) method developed by Richter et al. (1996) and the range of variability approach (RVA), which is a modification of the IHA (Richter et al., 1997). The IHA uses continuous stream flow data (either from stream gauges or generated by a model) to generate 32 hydrologic metrics characterizing the ecologically significant features of the flow regime. Hydrologic perturbations are evaluated by comparing "pre-impact" and "post-impact" time frames, using the central tendency and dispersion of each parameter to generate 64 IHA indicators. The IHA statistics consist of five groups:

- Magnitude of monthly water conditions;
- Magnitude and duration of annual extreme water conditions;
- Timing of annual extreme water conditions such as annual maxima and minima 1-day means or 30-day means;
- Frequency and duration of high (above the 75th percentile) and low (below the 25th percentile) pulses; and
- Rate and frequency of water condition changes.

A summary of the hydrologic parameters used in the IHA and their characteristics is given in Table 2-1. It is recommended that at least 20 years of record be used to differentiate between natural variation and regime alteration. In the RVA, the 32 hydrological metrics from the IHA method (Richter et al., 1996) are computed for every year of an unregulated daily streamflow record to characterize monthly flow variations, magnitude, frequency, timing of extreme events, and rise and fall rates of the hydrograph. The natural annual range of these metrics is identified and used to set acceptable ranges of flow to be used as management targets such as plus or minus one standard deviation, for example.

Table 2-1. Summary of Hydrologic Parameters Used in the IHA and Their Characteristics.

IHA Statistics	Regime Characteristics	Hydrologic Parameters
Group 1: Magnitude of monthly water conditions	Magnitude Timing	Mean value for each calendar month
Group 2: Magnitude and duration of annual extreme water conditions	Magnitude Duration	Annual maxima 1-day means Annual minima 1-day means Annual maxima 3-day means Annual minima 3-day means Annual maxima 7-day means Annual minima 7-day means Annual maxima 30-day means Annual minima 30-day means Annual maxima 90-day means Annual minima 90-day means
Group 3: Timing of annual extreme water conditions	Timing	Julian date of each annual 1-day maximum Julian date of each annual 1-day minimum
Group 4: Frequency and duration of high and low pulses	Magnitude Frequency Duration	No. of high pulses each year No. of low pulses each year Mean duration of high pulses within each year Mean duration of low pulses within each year
Group 5: Rage and frequency of water condition changes	Frequency Rate of change	Means of all positive differences between consecutive daily means Means of all negative differences between consecutive daily values No. of rises No. of falls

Source: Richter et al. (1996)

Suen et al. (2004) point out that a practical drawback of the IHA is that it is impossible to optimize all 32 metrics. Consequently, there have been some attempts to reduce the number of indices needed for IHA by identifying a subset of key metrics that are most ecologically important or that avoid redundancy in the information they provide (Clausen & Biggs, 1997; 2000; Olden & Poff, 2003). Clausen and Biggs (1997) attempted to determine the most ecologically relevant hydrologic indices for characterizing flow regimes by computing 34 hydrologic variables from daily flow records at 83 sites and testing them for statistically significant relationships with periphyton and invertebrates. They found that the frequency of flows greater than three times the median flow is the most useful in explaining the ecologic condition because it explains four out of six main benthic community measures. In a later study, Clausen and Biggs (2000) calculated 35 flow variables for 62 perennial rivers from a common 7-year flow record and identified four groups of variables based on covariance among sites using a principal component analysis (PCA). The highly correlated groups of variables were the following: 1) size of river (average flow magnitude); 2) overall variability of flow and magnitude of extreme events (intensity of disturbance); 3) volume and duration of high flows (bed movement and degree of disturbance); and 4) the frequency of high flow events.

Olden and Poff (2003) reviewed 171 hydrologic indices using long-term continuous flow records from 420 sites and used PCA along with ecologically based recommendations to select a reduced set of indices that sufficiently represent the critical attributes of the flow regime. The ecologically based recommendations included evaluation of ecologically important streamflow characteristics that constitute the natural flow regime, including the seasonal patterning of flows; timing of extreme flows; the frequency, predictability, and duration of floods, droughts, and intermittent flows; daily, seasonal, and annual flow variability; and rates of change (Poff et al., 1997). Olden and Poff (2003) recommended using a subset of two to four indices from 171 indices to adequately represent the biologically critical portions of the flow regime. They also examined whether hydrologic indices were transferable among different "stream types." The six different "stream types" were chosen to cover a range of regional flow regime types — harsh intermittent; intermittent flashy; snowmelt; snow and rain; superstable or stable groundwater; and perennial flashy. They found that indices explaining dominant patterns of variance were stream-type specific, demonstrating that hydroclimatic characteristics of a study region should be considered when choosing hydrologic indices for a study.

A number of researchers advocate the use of hydrologic metrics that quantify altered streamflow regimes to form mechanistic links between urbanization, hydrology, hydraulics, and biota (Scoggins, 2000; Kennen & Ayers, 2002; Eisele et al., 2003; Kirby, 2003; Booth et al., 2004; Cassin et al., 2005). McMahon et al. (2003) quantified the difference between flow-based and stage-based metrics in urban and non-urban catchments. The primary motivation for this comparison was that if stage-based metrics were found to be comparable to the widely used flow-based metrics, they could provide a cost-effective alternative hydrologic measure in catchments lacking flow-based metrics data. They found that stage-based measures of flashiness and median duration are most similar to the flow-based metrics, whereas low-flow metrics have the least similarity. In an urban versus non-urban catchment comparison, they found that hydrologic variability (flashiness) is positively correlated with urban intensity index, percentage developed land, population density, road density, and configuration of developed land patches and inversely correlated with forest cover. The results suggest that fragmentation (i.e., degree of scatter across basin) of developed land is negatively correlated with overall flow variability and flashiness (McMahon et al., 2003). These results are similar to the findings of Alberti et al. (2003) and provide a possible mechanistic explanation (flow variability and flashiness) for how land use patterns affect benthic communities.

Scoggins (2000) analyzed six years of rapid bioassessment (RBA) data from urban streams in the city of Austin, Texas, in conjunction with continuous flow data from USGS gages at the outlets of the catchments using IHA metrics along with statistical procedures recommended by Poff and Ward (1989). He demonstrated the usefulness of such procedures to differentiate between levels of development and geohydrological conditions and evaluate their relationship with stream biology. Ecologically relevant parameters such as rise rate, number of high pulses, and number of zero flow days were found to be much higher in the highly developed stream than in the undeveloped stream. The results of his regression analysis show a significant relationship between the biological metrics and the duration of high pulses, the date of the 1-day maximum flow, percentage of floods in a 60-day period, and the annual coefficient of variation.

Kennen and Ayers (2002) found that decreases in base flow, periodicity in peak discharge, and the flashiness of stream flow substantially affect the types and condition of invertebrate, fish, and algal communities in New Jersey streams, in part by their direct effects on the stream channel. Changes in mean annual peak flow are significantly related to all communities; however, the 2-year peak discharge was found to be significantly related only to the fish and algal community structure.

Kirby (2003) demonstrated linkages among hydrology, hydraulics, and stream biotic conditions by characterizing specific aspects of the flow regime altered by urbanization using the IHA method to characterize flow regimes along an urban gradient in the Potomac River basin. The in-stream hydraulics inferred from the IHA parameters were estimated using HEC-RAS. Finally, statistical relationships were established between specific hydrologic and hydraulic variables and macroinvertebrate community integrity. The results of a PCA identified altered flow predictability, low flows, and flashiness (rise and fall rates) as the most significant hydrologic parameters related to benthic macroinvertebrate community health. Analysis of the relationships between hydrologic change and modeled hydraulic factors indicated that hydraulic parameters including Froude number, measures of stream energy, channel width divided by drainage area, and cross-sectional flow area could be useful indicators of altered watershed condition if thresholds can be identified. Figure 2-1 illustrates the relative importance of the different classes of hydrologic variables to macroinvertebrate community integrity, based on the normalized number of statistically significant relationships observed in his study.

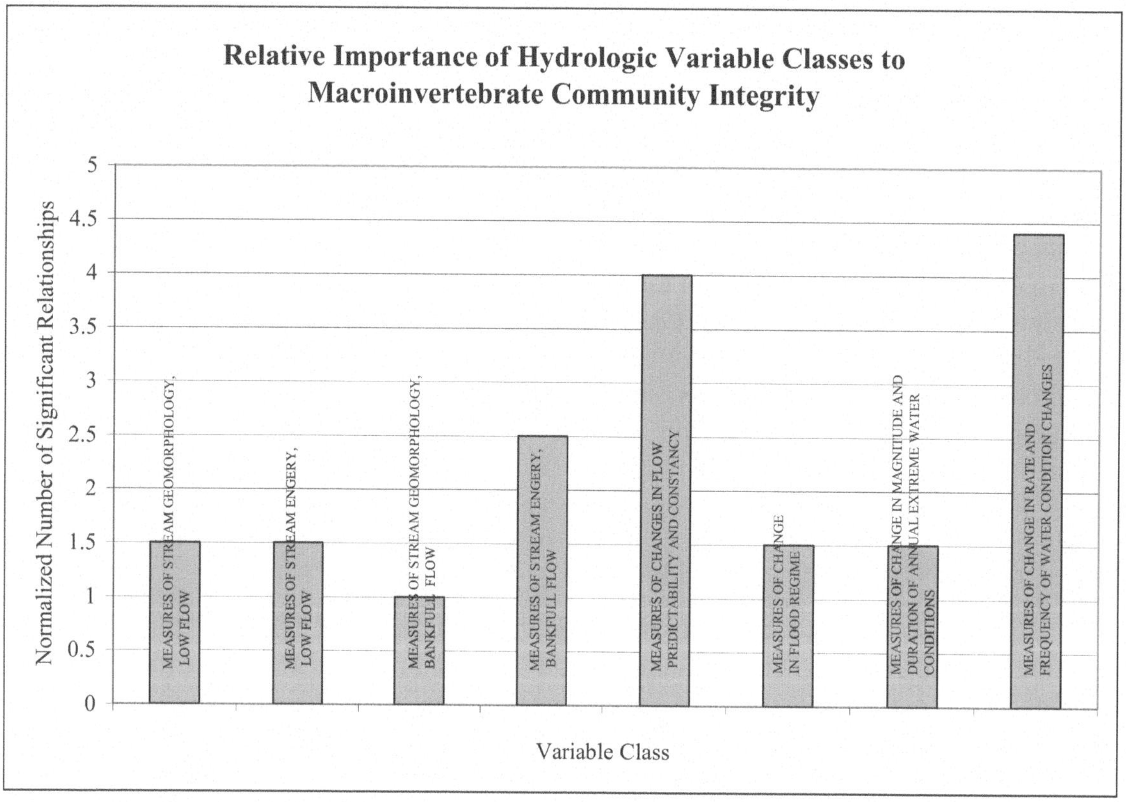

Figure 2-1. Variable Classes Ranked According to Their Relative Degree of Relatedness to the Measures of Biological Integrity (Kirby, 2003).

Stewardson and Gippel (2003) point out that a limitation of the RVA approach proposed by Richter et al. (1996; 1997; 1998) is that there is no specific way of relating ecologic response to the flow statistics used. They propose the events flow method (EFM) as a way to develop flow targets by explicitly identifying geomorphic and biological processes affected by flow variability. This is done by selecting threshold discharges, as well as event magnitudes and durations for events based on expert knowledge of important time scales for geomorphic and biological processes.

Konrad and Booth (2002) studied the effect of length of record on error in hydrologic trend analyses using hydrologic metrics derived from annual streamflow records to investigate the hydrologic trends associated with urbanization. They found that trend tests on any hydrologic metric using 10 years or fewer of records are likely to identify trends unrelated to urban development and fail to identify trends due to urbanization that act over periods longer than the period of record.

Booth et al. (2004) examined the relationship between land cover (TIA) and B-IBI in the Puget Sound Basin at three spatial scales: sub-basin, riparian (200-m wide buffer on each side of stream for the full length of the network upstream of from the sampling point), and local (200-m wide buffer for 1 kilometer upstream from the sampling point). They also used hydrologic metrics to characterize annual stormflow and baseflow patterns for a 12-year period. They chose to calculate the fraction of the year that the daily mean exceeds the annual mean, T_{Qmean}, as an indicator of flashiness (lower values meaning flashier). The fraction of a multiyear period that the flow exceeds the 0.5-year return period storm, $T_{0.5yr}$, was calculated as a measure of the influence of urbanization because high flows tend to increase in frequency, but not duration, as a result of development; that is, individual high flow events occur more often with more development. Figure 2-2 shows that flashiness (represented by two hydrologic metrics) increases with increased percentage impervious area. The relationship between these metrics and biotic integrity for the Puget Sound area is illustrated in Figure 2-3. It can be seen that B-IBI scores are higher in less flashy watersheds.

In the King County Normative Flow Project, Cassin et al. (2005) expanded the approach of Booth et al. (2004) to develop an ecologically based flow assessment method that could be used to evaluate the effects of past, present, and future water and land management actions on the flow regimes of streams and rivers; evaluate the effects of altered flow regimes on the physical and biological conditions in rivers and streams; and guide management actions to reduce these effects through mitigation, protection, and restoration of flow regimes that support ecosystem integrity and salmon conservation. Hydrologic metrics were computed from both stream gage data and flow data simulated by hydrologic simulation program Fortran (HSPF). Hydrologic metrics were evaluated to identify those that were not different between HSPF and stream gage data. Additionally, the metrics were also evaluated to identify those with the best correlations to benthic macroinvertebrate community data. The following hydrologic metrics were identified as appropriate for computation by modeling and as predictors of biotic response:

- Low-flow threshold pulse events and interval between pulses,
- High-flow threshold pulse events and total period of the year with high pulses,
- T_{Qmean},
- Percentage of time above the mean 2-year flow, and
- Timing of onset of fall flows.

As part of this protocol study, Edgerly (2006) and Edgerly et al. (2006) performed an investigation of hydrologic metrics to determine which might be most sensitive to changed land uses and stormwater runoff controls. Edgerly (2006) examined 10 hydrologic metrics (Table 2-2) proposed by various researchers to determine their sensitivity to urban runoff control strategies. The hydrologic metrics included: $T_{0.5}$, which is the percentage of time that the stream flow

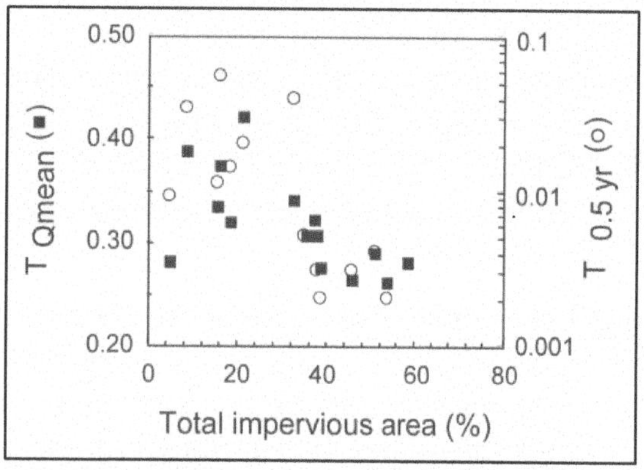

Figure 2-2. Discharge Flashiness as Measured by Two Hydrologic Metrics. T_{Qmean} on the Left Axis; $T_{0.5yr}$ on the Right Axis; Lower Values Are Flashier (Booth et al., *Journal of the American Water Resources Association,* 40(5): 1351–1364, 2004. Blackwell Publishing).

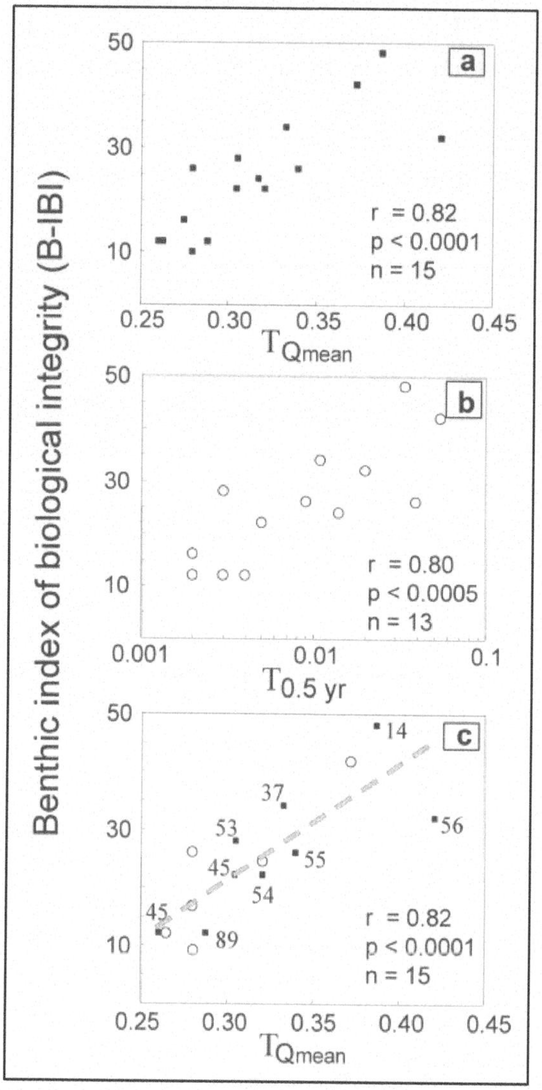

Figure 2-3. Relationship Between B-IBI and (a) T_{Qmean} and (b) $T_{0.5yr}$. In (c) Numbers Indicate Local Urban Land Cover Percentage. Sites Plotted as Circles Lack Local Land Cover Data (Booth et al., *Journal of the American Water Resources Association*, 40(5): 1351–1364, 2004. Blackwell Publishing).

exceeds the peak flow of the 0.5-yr storm; T_{Qmean}, which is the percentage of time that the stream flow exceeds the mean annual flow; high pulse duration rise rate, and fall rate metrics from Richter et al. (1996); event mean rise and fall rate; and the ratio of event peak duration to total event duration. These hydrologic metrics are described in Table 2-2. Of the 10 hydrologic metrics examined, Edgerly (2006) found that only the T_{Qmean} was sensitive to SWMM modeling of land use change. However, the T_{Qmean} was not found to be as sensitive to stormwater runoff control strategies as the $T_{0.5}$.

Table 2-2. **Summary of Hydrologic Metrics Evaluated by Edgerly (Reprinted with permission of the Nature Conservancy, from *Indicators of Hydrologic Alteration: User's Manual*, Version 7, July 2006).**

Annual Metrics	Description	Units
$T_{Q\ mean} = \left[\dfrac{\sum_{yr} t\left[\left(\dfrac{Q(t)}{Qmean}\right) > 1\right]}{1year}\right]$	The fraction of a year that the daily mean flow rate exceeds the annual mean flow rate for that year	Fraction of time
modified $T_{Q\ 0.5\ yr} = \left[\dfrac{\sum_{yr} t\left[\left(\dfrac{Q(t)}{Q_{0.5}yr}\right) > 1\right]}{1year}\right]$	Fraction of a year that the 15-min mean flow rate exceeds a fixed pre-development half-year peak flow rate.	Fraction of time
Event-based Metrics	**Description**	**Units**
Q_p = max Q(t) from t_i to t_f	Event peak flow rate	cms
$D = t_f - t_i + 1$	Event storm flow duration	hr
$t_p = t(Q_p) - t_i + 1$	Time to peak	hr
$\dfrac{t_p}{D} = \dfrac{t(Q_p) - t_i + 1}{t_f - t_i + 1}$	Ratio of time to peak to storm flow event duration	ratio
$R_N = \left(\dfrac{Q_p - Q_b}{t_p - t_i + 1}\right)$	Event mean rise rate	$\dfrac{cms}{hr}$
$F_N = \left(\dfrac{Q_b - Q_p}{t_f - t_p + 1}\right)$	Event mean fall rate	$\dfrac{cms}{hr}$
$D_{peak} = \sum_{t=t_i}^{t_f} t\left[\left(\dfrac{Q(t)}{0.75xQ_p}\right) > 1\right]$	Duration that the flow rate exceeds 75% of the peak flow rate in an event.	hr
$\dfrac{D_{peak}}{D}$	Ratio of event peak duration to total event duration	Ratio

Where:
t = time-step
$t(Q_p)$ = time-step of peak flow
t_i = start time-step of event
t_f = end time-step of event

Q(t) = flow rate at time = t
Q_p = peak flow rate of the event
Q_b = base flow rate
$Q_{0.5yr}$ = peak flow with 6-month return period
Q_{mean} = annual mean flow for that year

CHAPTER 3.0

GUIDELINES FOR APPLICATION OF THE PROTOCOL

In Chapter 1.0, the protocol was briefly described; Chapter 2.0 provided the literature background supporting the decision to relate the health of urban streams to development using hydrologic metrics such as $T_{0.5}$ and T_{Qmean} and to relate these hydrologic metrics to biologic metrics such as the B-IBI or Ephemeroptera, Plecoptera, and Trichoptera (EPT) richness. This chapter provides general guidance for application of the protocol to set stormwater management and design criteria for urban runoff control to maintain the desired level of biotic health in receiving streams in terms of the biologic metrics indicated above. Each step in the protocol is addressed in the following eight sections of this chapter. Section 3.6 provides applications of establishing relationships between land use practices, runoff control strategies, and hydrologic and geomorphic metrics.

3.1 Step 1 — Identify Resources for Protection/Preservation and Establish Goals

Urbanization and the hydrologic changes associated with increased impervious (paved) surfaces have degraded aquatic life to the extent that this pattern has been dubbed the "urban stream syndrome" (Walsh et al., 2005; Meyer et al., 2005). Monitoring to identify which factors are limiting to aquatic life is crucial to protection and restoration efforts. The selection of appropriate biotic goals (Step 1 of the protocol, Figure 3-1) and adequate monitoring capabilities to measure these goals accurately are critical to the entire process for the protection of urbanizing waters.

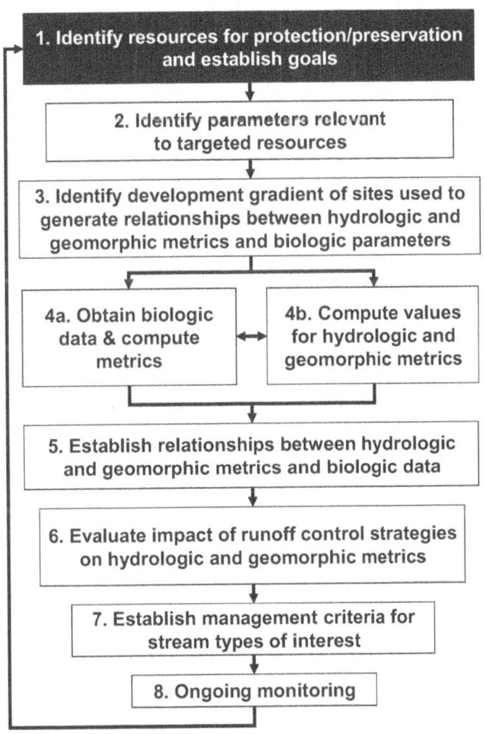

Figure 3-1. Identify Resources for Protection/ Preservation and Establish Goals.

The selection of biotic goals can be guided by existing state- or tribe-assigned aquatic life uses for a particular water resource. The basic regulatory impetus for protecting streams and rivers across the United States is the Federal Water Pollution Control Act or Clean Water Act of 1972 and amendments (33 U.S.C. §1251 et seq.; 40 C.F.R. parts 104-149), along with associated state water quality laws and rules that further define protection provided by each state. In addition, other laws such as the federal Endangered Species Act, or ESA (PL 93-205; 16 U.S.C. 1531 et seq.), and similar federal regulations and state laws protecting rare and endangered species can influence levels of protection provided to aquatic life in streams and rivers.

Integral to the selection of biotic goals is having adequate monitoring capabilities to measure attainment of these goals and to document likely causes of failure or non-attainment when goals are not achieved. The term "Data Quality Objective (DQO)" (see Section 5.1) has been used to refer to the process for identifying an objective of a study and then determining the data and its characteristics (e.g., accuracy, precision, etc.) that will provide a suitable answer to the objective. The U.S. EPA has recognized the importance of adequate monitoring and better integration of monitoring results into water quality management programs and is working to help define elements of a strong bioassessment program (U.S. EPA, 2002a and 2002b; 2003; Barbour & Yoder, 2005; Yoder, 2003). As called for in Step 2 of the protocol (Section 1), specific biological measures (e.g., the diversity or abundance of various elements of the fish or aquatic macroinvertebrate communities) must be identified for inclusion in monitoring efforts to determine attainment or nonattainment of biotic goals.

During selection of biotic goals and associated biological parameters/response indicators, a determination must be made whether the aquatic life uses for the study stream(s) are appropriate or whether they need to be adjusted via a use attainability analysis (UAA). Some states, such as Ohio, have well-documented UAA procedures and routinely apply these procedures to determine the appropriate use of previously unmonitored waters. For example, in the case of Ohio's UAA procedures, the Ohio EPA could assign a designated use of exceptional warm water habitat (EWH), warm water habitat (WWH), or modified warm water habitat (MWH) to a newly monitored stream, considering the condition of the stream. WERF has previously produced UAA guidance (Novotny et al., 1997) and is working to expand this in a recent request for proposals. If there is a consensus that aquatic life use designations are either too coarse or not linked directly to any biological measure, then investigators may need to consider developing more sensitive and appropriate goals for various waters. Even when a use change is not expected, a well-constructed UAA process can be used to select appropriate goals. Beyond the link to specific management goals, exploration of metrics or indices that are especially sensitive to urbanization stressors can be used to discriminate among the myriad of potentially limiting factors (Barbour et al., in press).

3.2 Step 2 - Identify Parameters Relevant to Targeted Resources

Step 2 of the protocol involves the identification of study parameters to be measured. These parameters may include measures of biological community condition, hydrologic and geomorphic parameters, chemical water quality analytes, and physical stream habitat measures. The selection of study parameters must correspond with the protection/preservation goals identified in Step 1. States and tribes might have existing study measures prescribed as part of their efforts to comply with the Clean Water Act requirements. Pursuant to Steps 2 and 4a of the protocol (Figure 3-2), biological parameters and sampling locations should be selected for and collected from a given study area based on the management goals that the study is intended to support. Whether selecting goals for urban waters or deciding which streams or sample locations to exclude or censor from further study, it is necessary to have a good conceptual model for the range of biological conditions expected in a region. Optimally, study sites should be not only exhibit the full range of development or land use in the study area, but also be representative of a biological condition gradient for aquatic life (U.S. EPA, draft 2005; Barbour & Yoder, 2005) in that region. Section 5.1.1 discusses the recommended process for determining a minimum set of biological parameters to include in a study of the biological condition of streams in developed and undeveloped watersheds. Identification of study parameters should be predicated on established data quality objectives (DQOs) that are ultimately tied to the intended use of the study and associated results.

Concerning hydrologic parameters, Table 3-1 summarizes a number of hydrologic

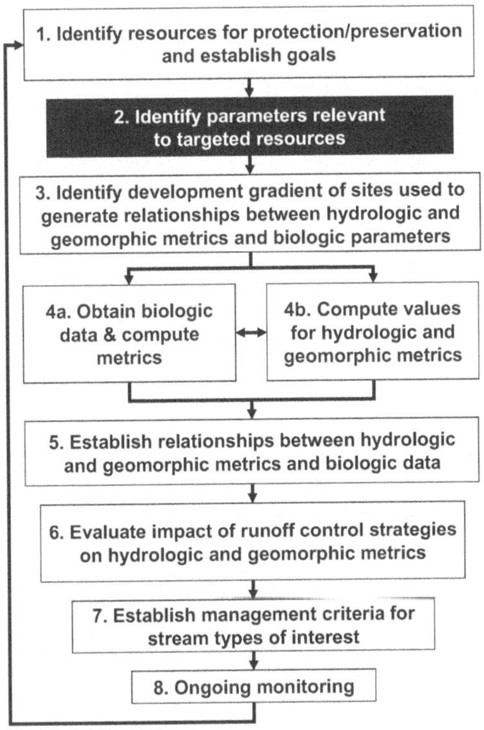

Figure 3-2. Identify Parameters Relevant to Targeted Resources.

metrics determined to be most useful in predicting biological condition in six different studies from five different regions. Table 3-1 also defines each of the hydrologic metrics.

Climatic region and stream type should be considered when determining which hydrologic metrics to evaluate. Olden and Poff (2003) presented a list of indices that best represent the biologically critical portions of the flow regime. These results are summarized in Table A-1 of the Appendix. Six different stream types are shown to cover a range of regional flow regime types, including harsh intermittent, intermittent flashy, snowmelt, snow and rain, superstable or stable groundwater, and perennial flashy. Tools for the identification of stream type are provided in Poff (1996). It should be noted that the metrics presented in Table A-1 have not been evaluated against biologic data to determine how effectively they predict

Table 3-1. Comparison of Hydrologic Metrics Found to be Most Useful for Predicting Biotic Integrity.

Study	Location	Measure of Biotic Integrity	Most Useful Hydrologic Metric for Predicting Biotic Integrity	Definition	Flow Regime Characteristic Represented by Metric
Clausen and Biggs (1997)	New Zealand	Periphyton	Q_{mean} or Q_{50}	Mean flow (m^3/s) or median flow (m^3/s), flow exceeded 50% of the time.	Absolute Size of River
		Invertebrates	FRE_3	Frequency of flows greater than three times the median flow.	Flow Variability and Flood Frequency
Scoggins (2000)	Austin, TX	Invertebrates	% of floods in a 60-day period	Common 60-day period for multi-year flow record that contains the largest percentage of floods (Poff and Ward, 1989).	Flood Predictability
			Mean Annual CV	The average over all years of the mean flow divided by the standard deviation times 100 (Poff and Ward, 1989).	Overall Flow Variability
			High Pulse Duration	Mean duration of high flood pulses, where high flood pulses are defined as the 75th percentile (Richter et al., 1996).	Flow Variability and Flashiness
			Date of High Pulse	The mean Julian date of the 1-day annual maximum flow over all years (Richter et al., 1996).	Timing/ Predictability
Kennen & Ayers (2002)	New Jersey	Invertebrates Fish Algae	RT2575	Ratio of discharge exceeded for the indicated percentage of time (Reed et al., 1997)	Flashiness
			2YRPK	2-year peak flow (Reed et al., 1997)	Flood Frequency
			10YRLOW	7-day, 10-year low flow (Reed et al., 1997)	Low Flows
Kirby (2003)	Virginia	Invertebrates	Rise/Fall Rates of Hydrograph	Means of all positive/negative differences between consecutive daily values (Richter et al., 1996).	Flashiness
			High Pulse Count	Number of high pulses within each year (Richter et al., 1996).	Flashiness
			Date of High Pulse	The mean Julian date of the 1-day annual maximum flow over all years (Richter et al., 1996).	Timing/Predictability
Booth et al. (2004)	Puget Sound, WA	Invertebrates	T_{Qmean}	The average annual fraction of a year that the daily mean flow exceeded the annual mean flow of the given year, which yields lower fractions for "flashy "streams and higher fractions for gradually varying flow regimes (Konrad and Booth, 2002).	Flow Variability and Flashiness
			$T_{0.5yr}$	The fraction of time that a stream channel is exposed to flows whose magnitude exceeds the half-year flood.	Flow Variability and Flashiness

Study	Location	Measure of Biotic Integrity	Most Useful Hydrologic Metric for Predicting Biotic Integrity	Definition	Flow Regime Characteristic Represented by Metric
Cassin et al. (2005)	King County, WA	Invertebrates	Low	Number of days between low pulse events, or the average number of days per occurrence that the hydrograph was below the low-flow threshold, for each calendar-year	Duration
			Low Pulse Range	The difference, in number of days, between the last and first times that the daily-time-step hydrograph crossed the low-flow threshold, for each calendar-year	Duration
			High Pulse Duration	Average number of days per occurrence that the hydrograph was above the high-flow threshold (200% of the average flow-rate under forested conditions over full period of record) for each water-year	Duration
			High Pulse Range	The difference in number of days between the last and first times that the daily-time-step hydrograph crossed the high-flow threshold (200% of the average flow rate under forested conditions over full period of record), for each water-year	Duration
			T_{Qmean}	The percentage of time (or days per year) in a given water-year that the daily-time-step hydrograph was above the year's average forested flow-rate (Booth and Konrad, 2002)	Flow Variability and Flashiness
			%of time above 2-year mean	The percentage of time that the daily or 15-minute time step was above the 2-year mean forested flow-rate	Flashiness
			onset of fall flows	The Julian date of the day after the annual 7-day minimum flow period for the year	Timing

biotic response; they are simply provided as a starting point to determine which metrics may be best at predicting biotic response.

In view of (a) the findings of Edgerly (2006) and Edgerly et al. (2006) that the T_{Qmean} was not found to be as sensitive to stormwater runoff control strategies as the $T_{0.5}$, (b) the results that were found using the $T_{0.5}$ in application of the protocol to the North Carolina watersheds (see Section 4.3), (c) the results obtained by Booth et al. (2004) in using this hydrologic metric in the examination of urbanized watersheds in the Pacific Northwest, and (d) Konrad's (2005) finding that unlike the T_{Qmean}, the $T_{0.5yr}$ was thought to be insensitive to basin size in Puget lowland stream basins, in application of the protocol it is recommended that the hydrologic metric $T_{0.5}$ be used to relate hydrologic metrics to benthic macroinvertebrate metrics such as EPT richness unless there are sufficient resources to examine and compare the viability of other hydrologic metrics.

These hydrologic metrics may be used as a point of departure in developing linkages among land use, hydrology, geomorphology, and biological condition. These hydrologic metrics are computed from a continuous flow record. If a continuous monitoring flow record of sufficient length is not available, the protocol requires that hydrologic and hydraulic models be developed for study watersheds to simulate a long-term continuous flow record for each. See Section 3.4.2 for a more detailed discussion of hydrologic sampling, model development, and metric development.

Erosion potential analysis can be used as a geomorphic metric to evaluate the impacts of urbanization and stormwater management practices. In addition, other aspects of stream substrates may be measured or modeled to evaluate the potential impacts of urban development. The substrate metrics could range from pebble counts to visual metrics such as the Ohio EPA qualitative habitat evaluation index (QHEI) embeddedness metric (Rankin, 1989). Critical shear stress values, estimated based on particle sizes or measured via in-situ jet-testing (Hanson and Cook, 2004) can be compared to calculated shear stress values using stream-specific erosion potential indices (Bledsoe, 2002). These indices could be utilized for the development of linkages among land use, hydrology, geomorphology, habitat, and biological condition. This information might also be utilized in the development of management criteria (Figure 3-2, Step 7). Erosion potential indices could also be calculated for watersheds targeted for protection in follow-up surveillance monitoring (Figure 3-2, Step 8) to evaluate the effectiveness of management efforts against stream channel erosion.

3.3 Step 3 – Identify Development Gradient of Sites Used to Generate Relationships Between Hydrologic and Geomorphic Metrics and Biological/Habitat Parameters

In Step 3 of the protocol (Figure 3-3) sites should be selected to represent a gradient of urbanization. The selected watersheds should represent a range of undeveloped, partially developed, and fully developed conditions as well as differences in density of development such as low, medium or high intensity land use. If flow monitoring data are used to apply the protocol, watersheds should be selected that have not had significant development or land use change during the monitoring period of record. If significant land use changes have occurred, a rainfall-runoff model can be used to develop a flow record that is based on existing land use in the watershed, rather than changes in land use over time.

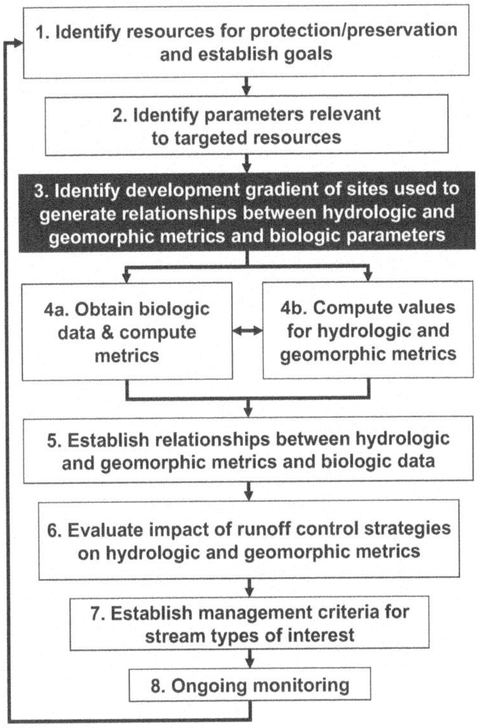

Figure 3-3. Identify Development Gradient of Sites.

For a regional study, at least 6 to 10 watersheds should be identified that meet the following criteria:

- **Homogeneous land use within each watershed:** The intent of this criterion is to maximize the potential for a study to identify impacts of different land use or development types. Urban and suburban watersheds commonly have a mix of low-, medium-, or high-density residential land uses intermixed with commercial land use, as well as urban open space such as parks, schoolyards, or cemeteries. Identification of homogenous land use in the watershed is not intended to select watersheds with the same land use throughout the watershed, rather, watersheds should be avoided that contain clustering of land use, which could include high-, medium-, or low-intensity residential development in the headwaters of the watershed and forest or agricultural land near the outlet of the watershed such as is shown in Figure 3-4a, or a split of urban land use along one tributary and no urban development along the other, such is shown in Figure 3-4b. Figures 3-4c and 3-4d show watersheds with a spatially distributed mix of land uses.
- **Similar stream geomorphic characteristics:** To the extent possible, streams in selected watersheds should exhibit similar geomorphic characteristics, regardless of the level of development in the watershed. Such characteristics should include substrate size, similar geomorphic channel units, and local geologic setting.
- **Watershed size and stream order:** Study watersheds generally should have an area no greater than approximately 5 to 10 square miles, with no larger than third-order streams. The rationale for this criterion is that larger streams tend to be somewhat buffered against the hydrologic impacts of development due to response of a more complex watershed system. Additionally, if development and stormwater management practices are not implemented uniformly throughout a watershed their impacts on hydrologic and geomorphic metrics may

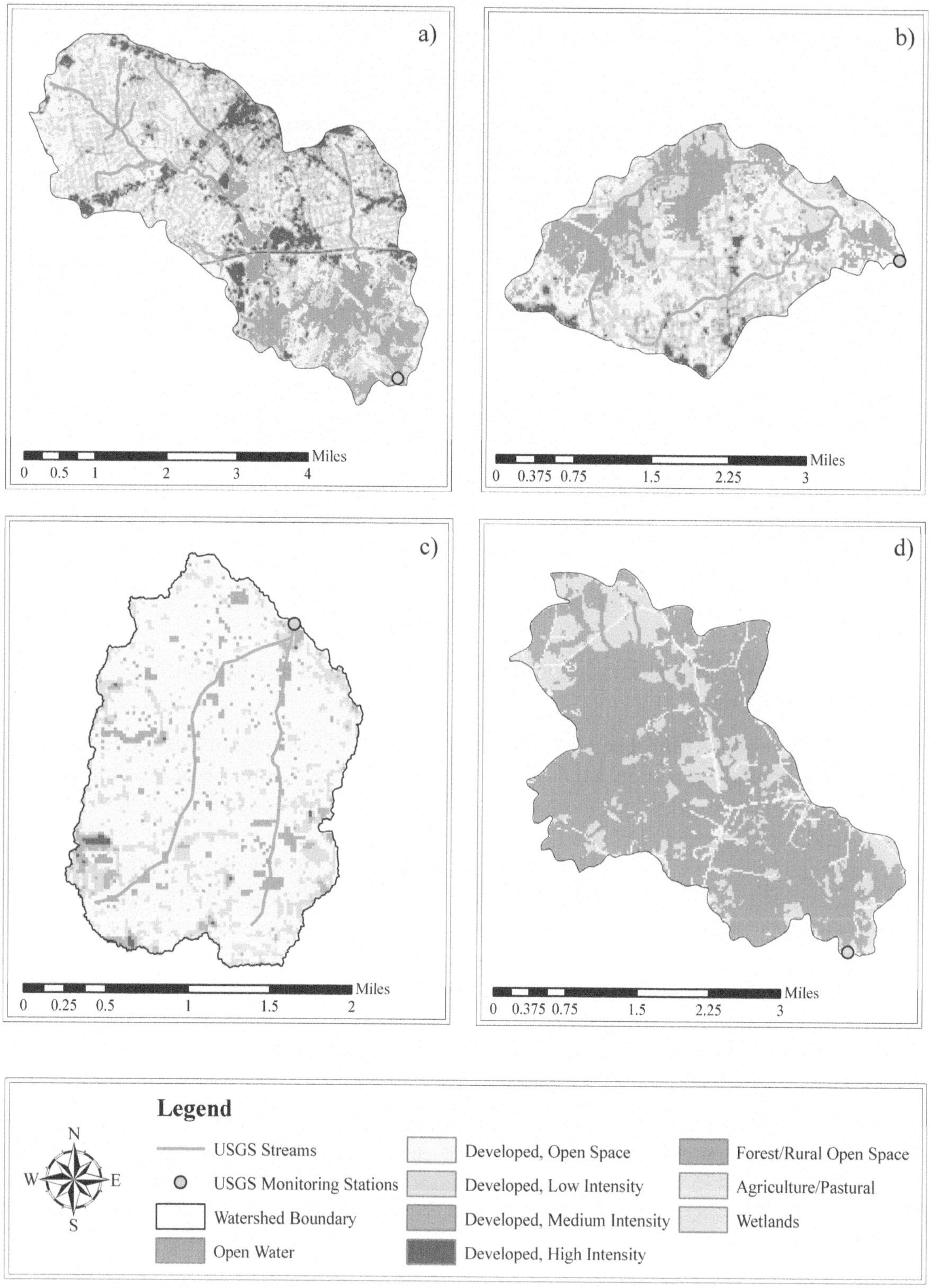

Figure 3-4. Examples of Land Use Distribution.

be difficult to discern. Streams with drainage areas of less than 10 square miles also tend to be wadeable streams that are accessible under normal flow conditions without the need to use a boat for field sampling. Wading methods may be used to sample streams with drainage areas as large as 150–200 square miles; however, sampling sites with such large upstream drainage areas require consideration of drainage area impacts related to changes in biotic assemblages.

- **Presence or absence of stormwater management practices:** Watersheds should be selected both with and without established stormwater best management practices (BMPs), particularly BMPs that improve water quality. The intent of this criterion is to enable studies to clarify the potential for attenuating the impacts of development with stormwater management approaches.
- **Avoidance of non-stormwater-related sources of potential impairment:** Watersheds with point source discharges, combined sewer overflows (CSOs), dams, transbasin diversions, and other significant river engineering activities should be avoided.
- **Flow gage availability:** Watersheds with flow gages should be preferred over watersheds without, with gages preferably in close proximity to biological sampling stations. The intent of this criterion is to support calibration of hydrologic models to simulate flows at biological sampling stations.
- **Rain gage availability:** Watersheds with rain gages should be preferred over watersheds without gages, in support of hydrologic modeling potential.
- **Elimination of acutely impaired sites:** Study sites should be screened to avoid including acutely impaired sites that exhibit water quality that is sufficiently poor to potentially obscure the effects of other development-related stressors. Details on water quality parameters for screening are given in Section 5.3.

Sites should be selected that represent a gradient of urbanization from undeveloped to full build-out. The extent of urbanization can be determined using many factors such as gross impervious area (GIA), directly connected impervious area (DCIA), population, housing unit or road density, percent of urban land cover or a combination of these factors. The urban intensity index (UII) described by McMahon and Cuffney (2000) combines these factors to compare the extent of urbanization of watersheds in urban gradient studies conducted by the USGS. Factors describing the extent of urbanization in watersheds can be used as tools for the preliminary identification of candidate watersheds for monitoring. Further screening of these candidate watersheds is required with respect to geomorphic and land use homogeneity of the watersheds. Sampling sites that are typical of streams in the area, with suitable locations for flow measurement available close to the biologic/water quality sampling sites, should also be present for the watersheds selected. If flow data are not available, simulation of a continuous flow record requires data from continuously recording rain gages with at least 5 years of hourly or 15-minute (preferred) data.

If it is not economically or physically feasible to study a gradient of urbanization, paired watersheds or a single watershed may be used to apply the protocol. Hydrologic and geomorphic metrics can be developed for a single, undeveloped watershed and development scenarios could be tested using hydrologic and hydraulic models to identify management practices that closely reproduce the desired hydrologic flow regime. Similarly, a pairing of undeveloped and developed watersheds can be used to identify targets for rehabilitation of a developed and degraded watershed. It is important to note, however, that there is a level of uncertainty associated with the computed hydrologic and geomorphic metric values, and that use of one data point as a representative target does not allow for a range of target values to be identified.

3.4 Step 4 — Development of Data to Support Linkages

3.4.1 Step 4a — Obtain Biological Data for Each Site and Compute Biologic Metrics

Pursuant to Steps 2 and 4b of the protocol (Figure 3-5), biological parameters and sampling locations should be selected for and collected from a given study area based on the management goals that the study is intended to support. The biologic metrics selected will be regional or state dependent. Both selecting goals for urban waters and deciding which streams or sample locations to exclude or censor from further study require a good conceptual model for the range of biological conditions that can be expected in a region. As noted under Step 3, study sites should be selected that not only exhibit the full range of development or land use in the study area but that are also representative of a biological condition gradient for aquatic life (U.S. EPA, 2005; Barbour and Yoder, 2005) in that region. Details on the recommended process for determining biological and habitat parameters for support of the protocol are given in Chapter 5.0. If it is unclear which biologic index is the best to use, it is suggested that the data collection program include, at a minimum, the data required to compute metrics for benthic macroinvertebrates such as EPT richness or the B-IBI. The biologic data must be synthesized into indices that will be used as measures of the biotic health of streams draining each watershed. The procedure for computing biologic indices is regionally dependent and will not be discussed further here. It is recommended that two or more indices be tested.

3.4.2 Step 4b — Compute Values for Hydrologic and Geomorphic Metrics for Each Site

Step 4b involves working with a continuous flow record to compute hydrologic and geomorphic metrics. If a continuous flow monitoring record of *at least* 5 years of hourly or 15-minute data is not available for each sampling site, the data set can be extrapolated using computer simulation models to represent the complete hydrologic and geomorphic character of the watersheds being studied. If flow monitoring is less than 5 years, hydrologic metrics computed from that short hydrologic record will not reflect the true value of the hydrologic metric because it will not reflect the full range of storms that occur in the watershed. Moreover, the biologic data likely reflect the hydrology of the last 3 to 5 years, (See Section 3.3) so it is recommended that the hydrologic metric data be computed using a 5-year record, unless that record includes a very large storm or a severe drought. In addition, an even longer period of record is recommended for computation of geomorphic impacts, so that the effects of large (> 5-year return period) storms can be used to evaluate changes in the potential for erosion on the stream channel. Though it is generally believed that the channel-forming discharge occurs between a 1- to 2-year return period, the effects of larger events are not inconsequential and should be considered when evaluating geomorphic metric values.

3.4.2.1 Development of Continuous Simulation Models for Computing Hydrologic and Geomorphic Metrics and Examining Impacts of Alternative Land Use Practices and Runoff Control Scenarios

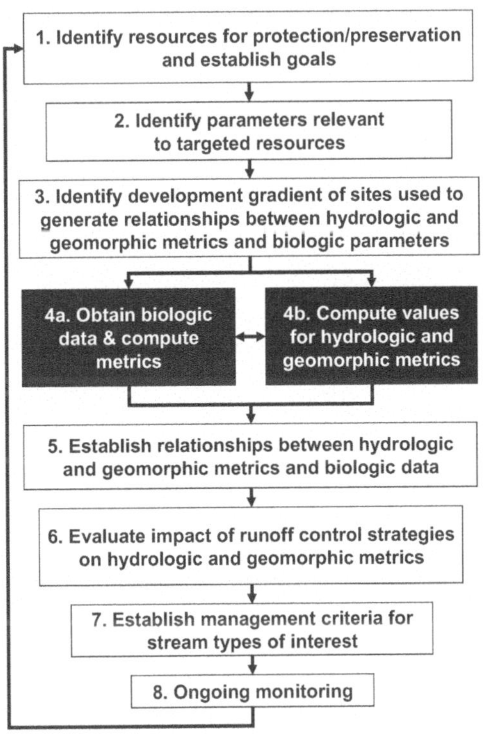

Figure 3.-5. Development of Data to Support Linkages.

There are a variety of hydrologic and hydraulic mathematical models that can be used for continuous simulation of stream flow in watersheds. Two of the most popular are HSPF and EPA's SWMM. Other models can be applied, but they must meet the following criteria:

- Be able to process a continuous 20- to 50-year record of 15-minute rainfall data and translate it to runoff accounting for land use practices within the watershed;
- Simulate infiltration processes that may vary across the watershed;
- Simulate subsurface flow;
- Route flows dynamically through the stream channel drainage system;
- Simulate the effect of runoff controls, such as BMPs and detention storage, within subcatchments, on the runoff hydrographs from the subcatchments; and.
- Be able to provide a time series of channel hydraulic properties (at 15-minute intervals), including stage and discharge, during runoff and dry weather periods.

EPA SWMM5 is well suited to these criteria; therefore it was used in the application of the protocol to the Raleigh-Durham, NC area. Use of EPASWMM5 for application of the protocol is explained in the remainder of this section. For guidance in using other computer models, please refer to the appropriate guidance documentation for that model.

Figure 3-6 is a logic diagram that can be used to assist a researcher applying a hydrologic and hydraulic model to a watershed. Note that under certain conditions it may not be possible to apply the model to the desired watershed in which case that site cannot be used to relate biologic metrics to hydrologic metrics for your study area. This situation would occur if you do not have a

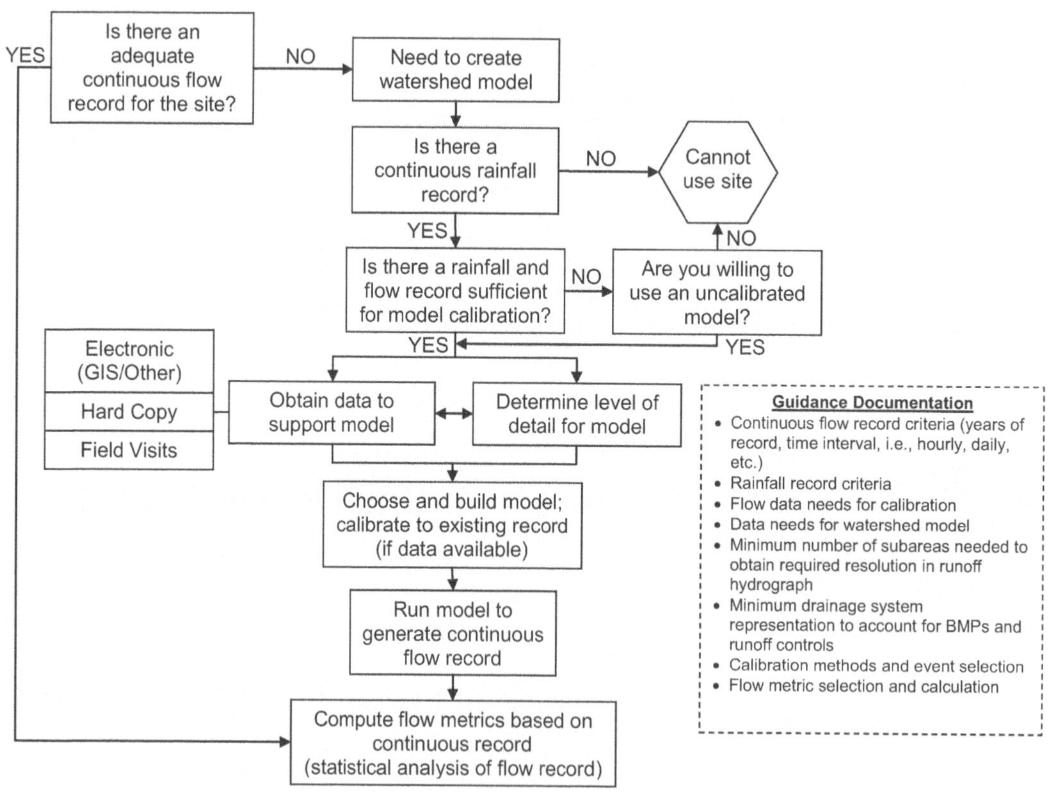

Figure 3-6. Logic Diagram for Applying a Simulation Model for Computing Flow Metrics.

continuous rainfall of sufficient duration (at least 5 years) or you do not have sufficient synoptic measurements of rainfall and stream flow to calibrate your model and you are unwilling to use an uncalibrated model.

The data required for watershed modeling in SWMM5 is illustrated in Figure 3-7. Additional guidance can be found in the *SWMM5 Documentation Manual* (Rossman, 2005) or in Huber and Dickinson (1998). Data gathering and collection for input into the simulation model is not a trivial task. Investigators should be careful to allocate sufficient time for computer modelers to set up and calibrate the models on the test watersheds because the model output is a key decision variable in determining the stormwater runoff control criteria for protection of urban streams. In this study, approximately 18 staff-months of effort were used developing, calibrating and running the simulation models on the eight watersheds analyzed in the North Carolina Pilot Study.

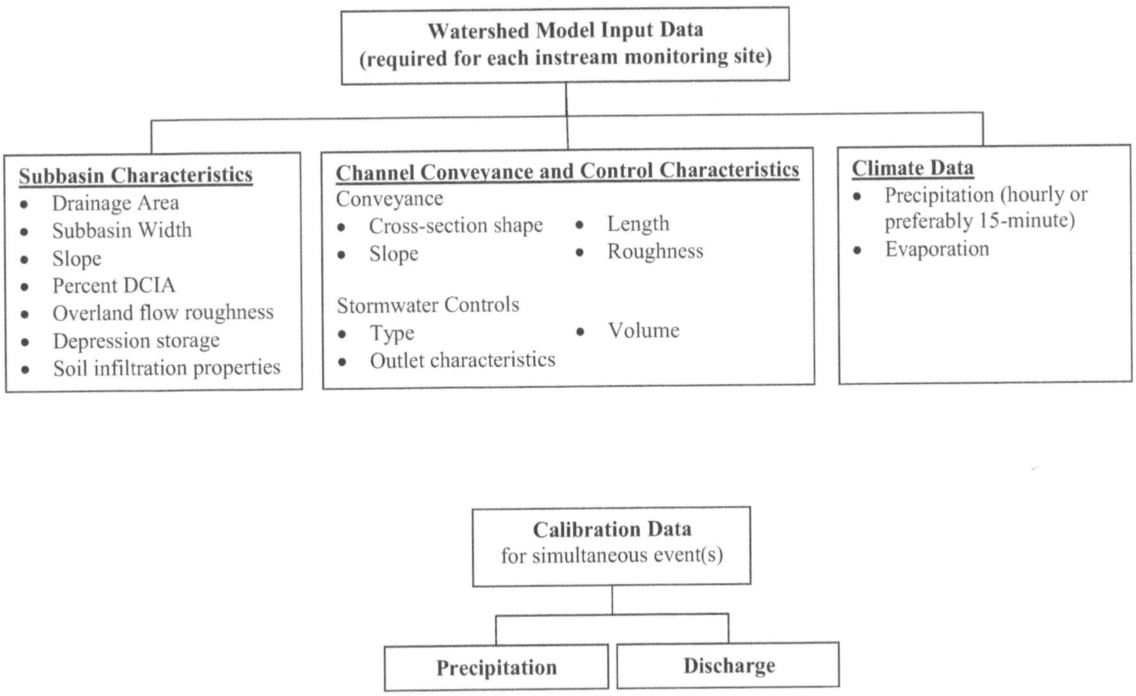

Figure 3-7. Data Requirements for SWMM5 Modeling.

SWMM and other hydrologic and hydraulic models have the ability to evaluate the hydrologic impacts of land use change and stormwater management practices. However, these models should be used with caution for the computation of hydrologic and geomorphic metrics of biologic interest. Many factors influencing historical or natural flow regimes (i.e., large woody debris, beavers, wetlands) as well as factors influencing current and future flow (i.e., influence of groundwater withdrawls on summer baseflows) may not be accurately estimated by these models (Cassin et al., 2005).

3.4.2.2 Calibration of Continuous Simulation Models

Calibration is the adjustment of watershed parameters in a model to generate a modeled behavior that best matches the monitored behavior of a watershed over a variety of runoff events. In this case, the process implied adjusting model parameters (such as infiltration rates and imperviousness levels) to obtain modeled results that match or approximate measured discharge records. Calibration can be done at a number of temporal scales (see Figure 3-8):

- Single event: one storm.
- Multiple events: several storms.
- Continuous simulation: long records (months to several years).

Calibration to a single storm is not appropriate for continuous simulation modeling because adjustment of variables to match runoff from one event may over or under adjust variables and inhibit matching of other events. It is recommended that a minimum of three to five storms be used for calibration. One or two of the storms should be small so that the only runoff that occurs is off the impervious areas. The other two or three storm should be large enough so that runoff occurs from the pervious area also but not so large that flooding occurs in the watershed. Additionally, these storms should occur after at least 7 days of no rainfall so that the soil is dry and the maximum infiltration rate can be estimated.

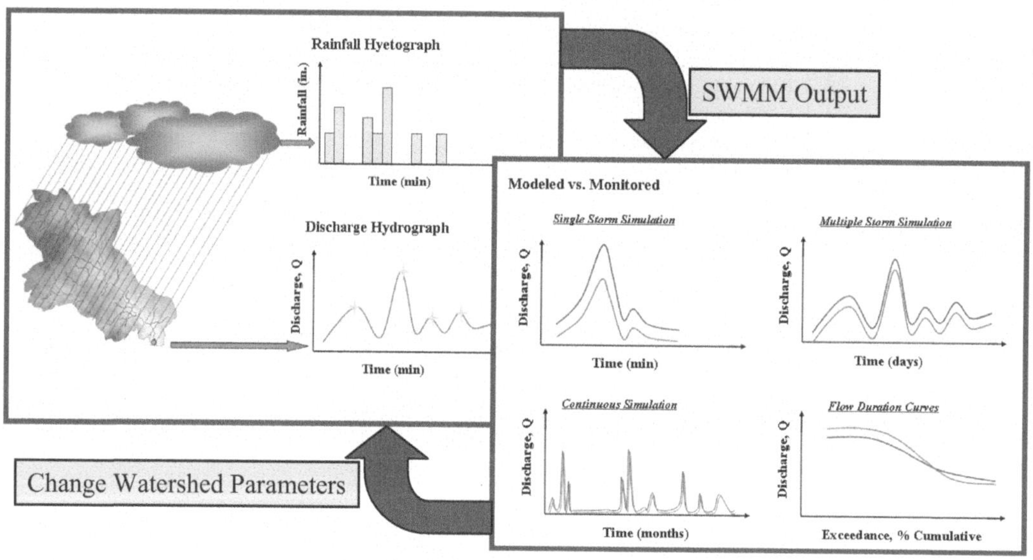

Figure 3-8. Calibration at a Variety of Temporal Scales.

If the rain gage that will be used to drive the simulation model is more than a mile away from the centroid of the watershed, it is recommended that an initial calibration be done using five discrete storms, but that the final calibration be conducted by comparing peak-flow frequency exceedance and flow-duration curves from the model to those developed from measured data. The idea is that the rainfall on the watershed might not correspond exactly to the rainfall measured at the gage because of the distance from the gage. Over a longer period of time, the rainfall frequency distribution on the watershed being simulated is likely to be similar to the rainfall frequency distribution at the gage supplying the rainfall data for the model, so the simulated and measured peak-flow frequency and flow-duration curves should be similar.

The effect of snowmelt is also an important factor to consider for model calibration. Although flow rates are typically low, they may be sustained over several days. Rainfall events superimposed upon snowmelt baseflow may produce higher runoff peaks and volumes as well as add to the melt rate of the snow. Both EPASWMM and HSPF have the ability to perform snowmelt computations while running continuous simulations (Huber and Dickenson, 1998; Bicknell et al., 1997).

3.4.2.3 Computation of Hydrologic Metrics

Hydrologic metric data must also be developed from the flow record measured prior to and during the biologic/water quality data collection program or from continuous computer simulation models. Under Step 2, a number of different hydrologic and geomorphic metrics were identified, which various investigators have determined relate to stream biotic health.

Edgerly (2006) and Edgerly et. al. (2006) found that that stream hydrologic metrics have traditionally been computed using daily mean values of flow. Although this method of computation is valid for large watersheds, the rainfall-runoff response of watersheds recommended for this protocol (less than 10 square miles) is so rapid that trends in hydrologic metrics are not easily detected using a mean daily flow time series. To make the statistics meaningful for watersheds of this size, it is recommended that hydrologic metrics be computed from 15-minute or hourly flow data because the streams in the size watersheds recommended in Step 2 for sampling will respond to these flow changes over these short time intervals. If watersheds used in the protocol are smaller than 3 square miles, investigators should evaluate the sensitivity of the metrics computed from data at both hourly and 15-minute resolution to determine if hourly data are adequate or if 15-minute data are required.

Computation of hydrologic metrics is best performed using a program capable of handling large data sets such as SAS or MATLAB applications. Several programs currently exist for the computation of hydrologic metrics: the National Hydrologic Assessment Tool (NATHAT) (Henrickson et al., 2006), the IHA (The Nature Conservancy, 2007), and GeoTool (Bledsoe et al., in press). NATHAT and the IHA software are only able to work with daily streamflow data. GeoTool is able to process data at various temporal resolutions; however, the hydrologic metrics are computed at daily increments only. Unfortunately, a daily resolution is not adequate to represent changes in land use or stormwater management practices in watersheds of the size recommended in this protocol, therefore these programs are not applicable to the computation of hydrologic metrics at 15-minute or hourly time steps.

3.4.2.4 Computation of Geomorphic Metrics

Excess shear stress and erosion potential indices are geomorphic metrics that can be used to evaluate the impacts of urbanization on stream channels. Shear stress duration curves can be used to help evaluate the full range of shear stress forces exerted on a stream channel, such as the one shown in Figure 3-9 for Cedar Creek in Lenexa, Kansas. The duration curve is generated by calculating average boundary shear stress in the stream channel for each time step in the measured or modeled flow record. Average boundary shear stress, τ_o is calculated as:

$$\tau_o = \gamma R_h S_f \qquad (3.4.2.4\text{-}1)$$

where, γ is the unit weight of water, R_h is the hydraulic radius of channel, and S_f is the energy gradient in the channel.

The potential for erosion in a channel can be quantitatively evaluated as the difference between *calculated* shear stress and *critical* shear stress. An instream erosion potential index can be used to estimate an increase or decrease in erosion potential by calculating a ratio of post-development to pre-development "work" on a channel (Bledsoe, 2002). This "work" on a channel can be estimated using erosion rates. The erosion potential index is calculated as a ratio of the sum of excess shear (ε) for each flow record in a continuous time series. Similar to computation of hydrologic metrics, the flow record can come from monitoring data or from the output from a continuous simulation model. The erosion potential index uses a simple finite-difference approximation to estimate the time-integrated sediment transport capacity over the duration of the continuous flow record:

$$E = \frac{\sum_{t=0}^{T} \varepsilon_{post}}{\sum_{t=0}^{T} \varepsilon_{pre}} \tag{3.4.2.4-2}$$

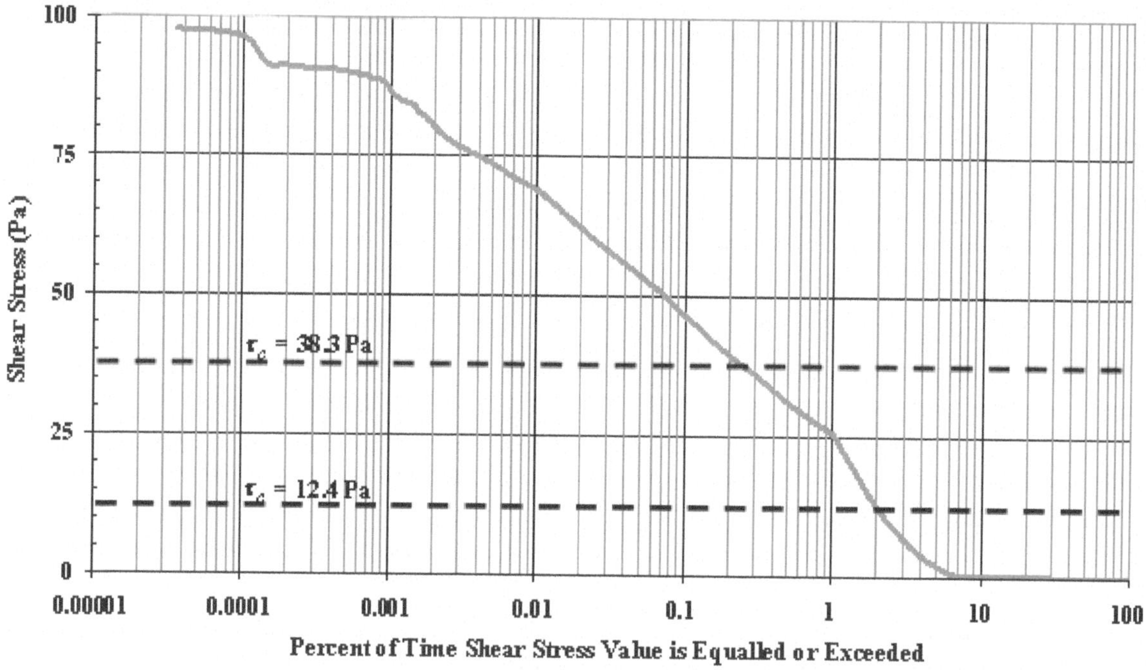

Figure 3-9. Shear Stress Duration Curve for Cedar Creek in Lenexa, Kansas.

where E is the instream erosion potential; ε is the erosion rate at time t; T is the length of the continuous flow record; and *pre* and *post* represent pre- and post-development conditions, respectively.

Channel erosion is a function of the erodibility of the bed and bank materials and excess shear stress. Erosion rates can be estimated using an excess shear equation described by Foster et. al. (1977):

$$\varepsilon = k\left(\tau_o - \tau_c\right)^A \tag{3.4.2.4-3}$$

where ε is the erosion rate, k is an erodibility or detachment coefficient, τ_c is the critical shear stress, and A is an exponent, which is typically assumed to be 1.5. When performing erosion potential analysis, the value of k is factored out and can be assigned a value of 1.0 to simplify calculations. For granular (non-cohesive) materials, τ_c can be estimated using the Shields criteria:

$$\tau_{*c} = \tau_c / \left(\gamma_s - \gamma\right) d \tag{3.4.2.4-4}$$

where τ_{*c} is the critical dimensionless shear stress, γ_s is the unit weight of sediment, γ is the unit weight of water, and d is the representative particle diameter.

Critical shear stress values can be used to identify the portion of a shear stress duration curve that must be matched to minimize increases in erosion potential in a stream channel, as shown in Figure 3-9. Approximate threshold values of τ_c for granular material at 20°C are given in Table 3-2. It should be noted that the values in Table 3-2 are based upon the classical relations for the threshold of motion of uniform sediment presented by Shields (1936), with an average τ_{*c} of 0.047. Parker (in press) suggests that the data presented by Shields (1936) tends to overpredict the critical dimensionless shear stress value and that this value is more appropriately 0.03. Given the uncertainty associated with hydrologic and hydraulic modeling and the overall variability of stream channel substrate, use of the values in Table 3-2 are adequate for computation of erosion potential until revised values of critical shear stress become widely available for geomorphic computations.

It is more difficult to estimate the erosion resistance of cohesive materials. Various studies on the erodibility of cohesive materials have shown that numerous soil properties influence erosion resistance, including antecedent moisture, clay mineralology and proportion, density, soil structure, organic content, and pore and water chemistry (Grissinger, 1982). Erodibility coefficients and critical shear stress values for cohesive sediments can be estimated by in-situ jet-testing using the methods described by Hanson and Cook (2004).

Erosion potential is a geomorphic metric that takes into account the effects of changes in flow regime of a stream channel, but it does not account for the effects of sediment supply or other factors likely to affect stream stability. Different types of urban streams are likely to exhibit varying degrees of instability (depending on entrenchment), relative erodibility of bed and banks, riparian condition, mode of sediment transport (bedload versus suspended load) and proximity to geomorphic thresholds (Bledsoe and Watson, 2001). Changes in land use (i.e., from agricultural to residential) combined with the sediment trapping efficiency of structural BMPs, such as extended detention ponds, may reduce bed material loads from the watershed. This reduction in bed material load, especially when combined with changes in flow regimes (as described by the erosion potential metric), can exacerbate channel instability, especially in mixed- or suspended-load channels. Additionally, a reduction in bed material load from the watershed means that reproducing the predevelopment flow regime may not maintain the stability of the stream channel, especially for those comprised of finer grained materials such as sand (Bledsoe, 2002).

Table 3-2. Approximate Threshold Conditions for Granular Material at 20°C. (After Julien, 1998. Reprinted with permission of Cambridge University Press).

Class Name	d_s (mm)	τ_c (Pa)
Boulder		
Very large	>2,048	1,790
Large	>1,024	895
Medium	>512	447
Small	>256	223
Cobble		
Large	>128	111
Small	>64	53
Gravel		
Very coarse	>32	26
Coarse	>16	12
Medium	>8	5.7
Fine	>4	2.71
Very Fine	>2	1.26
Sand		
Very Coarse	>1	0.47
Coarse	>0.5	0.27
Medium	>0.25	0.194
Fine	>0.125	0.145
Very Fine	>0.0625	0.110
Silt		
Coarse	>0.031	0.083
Medium	>0.016	0.065

3.5 Step 5 — Establish Relationships between Hydrologic and Geomorphic Metrics and Biologic Data

Relationships between the hydrologic and geomorphic metrics and biologic and habitat parameters are established during Step 5 of the protocol (Figure 3-10).

Methods for developing these relationships include regression, logistic regression, classification and regression tree (CART) analysis such as used in WERF IIA process (Paulson et al., 2001) or using Bayesian networks (Borsuk et al., 2004).

The simplest method for developing linkages between hydrologic and geomorphic metrics and biologic parameters is by simple regression. Regression determines the form and strength of a relationship between two variables; the intent is to predict a value of a dependent variable from a given independent or predictor variable (Dytham, 1999). In this protocol, the dependent variables are the biologic metrics and the independent variables are the hydrologic and geomorphic metrics. Regression

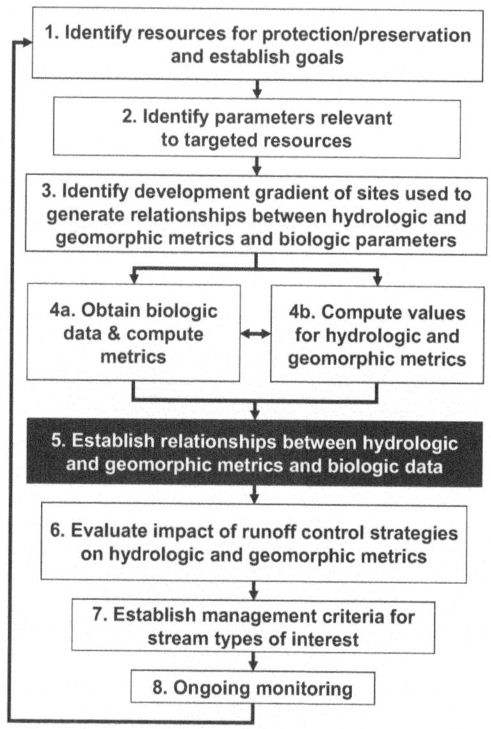

Figure 3-10. Establish Relationships Between Metrics and Data.

is easily applied and understood and was recommended by the project steering committee (PSC) as the method most appropriate for use by WERF Subscribers. Additionally, regression has been proven useful in relating biologic metrics to hydrologic metrics in studies throughout the United States (Cassin et al., 2005; Booth et al., 2004; Scoggins, 2000).

Sample size should be taken into account when selecting the number of metrics to evaluate. Fewer sites will mean a higher level of uncertainty in the results. A general rule-of-thumb for building and evaluating statistical models is to have 10 sites for each variable evaluated; therefore, a recommendation of 6–10 sites for this protocol was made in Section 3.3, keeping in mind the potential resource limitations of WERF Subscribers. If more sites are included in the protocol analysis, additional variables can be examined. Additional sites could also be used for cross-validation or testing of the relationships developed. At a minimum, relationships between causal variables should be examined, and users should keep in mind that relationships between hydrologic and geomorphic metrics are confounded by other factors, such as water quality, and that the metrics described in this report may not be the sole drivers behind the biologic results.

3.6 Step 6 — Evaluate Impact of Runoff Control Strategies on Hydrologic and Geomorphic Metrics

3.6.1 General Guidance

Step 6 is the Planning/Analysis task (Figure 3-11). Here, various land use and stormwater management scenarios are created and tested to determine their effects on hydrologic and geomorphic metrics. Examples of the stormwater management scenarios that can be evaluated include use of low-impact development practices or traditional BMPs such as extended detention ponds, swales, or constructed wetlands. The basic algorithm for this testing is illustrated in Figure 3-12. This testing is conducted using hydrologic and hydraulic models with continuous precipitation records to generate a continuous flow record. Resulting metrics computed from the continuous flow record for the various management scenarios are then compared against metric targets to determine which scenarios most closely match targeted goals. This process is part of Step 7 of the protocol which provides guidelines for the establishment of management criteria.

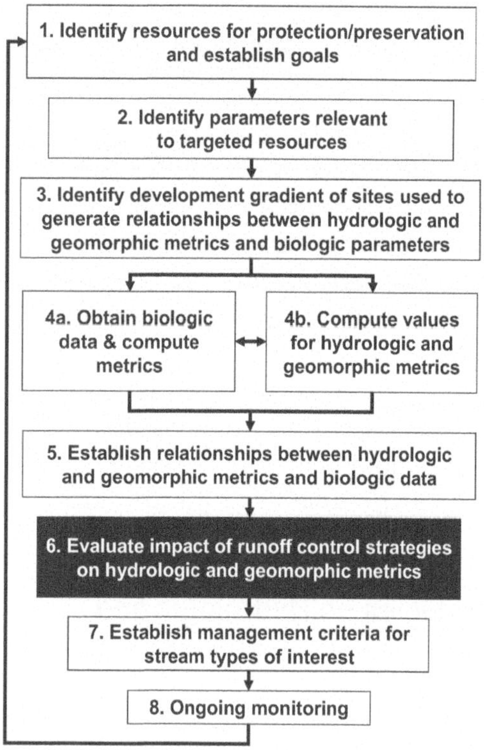

Figure 3-11. Establish Relationships Between Management Practices and Metrics.

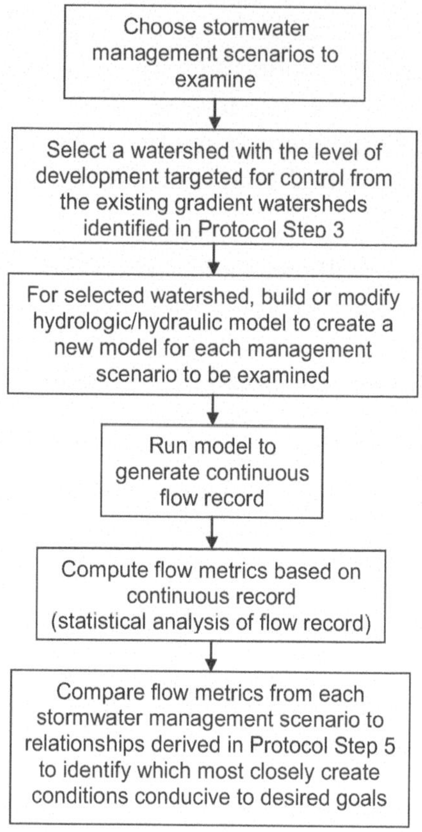

Figure 3-12. Basic Algorithm for Establishing Relationships Between Land Use Practices and Hydrogeomorphic Metrics.

3.6.2 Examples of Application

3.6.2.1 Impact of Runoff Control Strategies on Hydrologic and Geomorphic Metrics in a Hypothetical Development in Fort Collins, Colorado

Hydrologic and hydraulic modeling capabilities in EPA SWMM4.4h were used to evaluate the effectiveness of typical stormwater management practices on hydrologic and geomorphic metrics in a 10-ha watershed in Fort Collins, Colorado (Rohrer, 2004; Rohrer & Roesner, 2006; Edgerly et al., 2006). Fifty-year continuous simulations with hourly rainfall records from 1951 to 2000 were used to produce peak flow frequency exceedance and flow duration curves as well as to calculate the $T_{0.5}$ hydrologic metric. Stream channel erosion potential rates for a variety of critical shear stress values representative of both cohesive and non-cohesive sediments were also examined using the continuous flow record. Six different land use and runoff control scenarios were evaluated and are described as follows:

- The ***Undeveloped*** watershed was simulated as a parcel of pastureland with 5% impervious area where runoff was modeled to travel as overland flow to the outlet (Figure 3-13a).
- The ***Developed Uncontrolled*** watershed was modeled as a medium-density residential development with an average of 30% impervious area and a system of gutters, pipes, and swales directly connected to the outlet (Figure 3-13b). No stormwater controls were applied to the watershed in this scenario.
- In the ***Over Control*** scenario, runoff from the medium-density residential development described above was routed through controls designed to reduce the post-development 100-

year peak flow rate to the pre-development 2-year flow rate. This practice is a stormwater management requirement for new development in Fort Collins and is intended to compensate for high-impact development that has already occurred by "over-controlling" runoff from all new developments.

- In the ***Over Control + Best Management Practices (BMP)*** scenario, over-control detention is combined with a water quality control BMP designed to capture 70% to 90% of all runoff-producing events and have a draw-down time of 24 hours. These structures are designed to capture the "first flush" of runoff, which typically has the highest concentrations of pollutants. The volume necessary to capture the first flush is referred to as the water quality capture volume (WQCV). The WQCV is intended to allow time for pollutants contained in the runoff to settle out before the detained water is discharged to the stream. The WQCV was calculated to be 0.041 ha-m, using the Denver Urban Drainage and Flood Control District (UDFCD) (2001) procedure.
- Two ***Peak Shaving + BMP*** control scenarios were evaluated using the standard practice of reducing the post-development peak flow rates to pre-development flow rates: (1) peak shaving of the 100-year storm and (2) peak shaving of the 100- and 10-year storms. Both of these scenarios were combined with a water quality BMP.

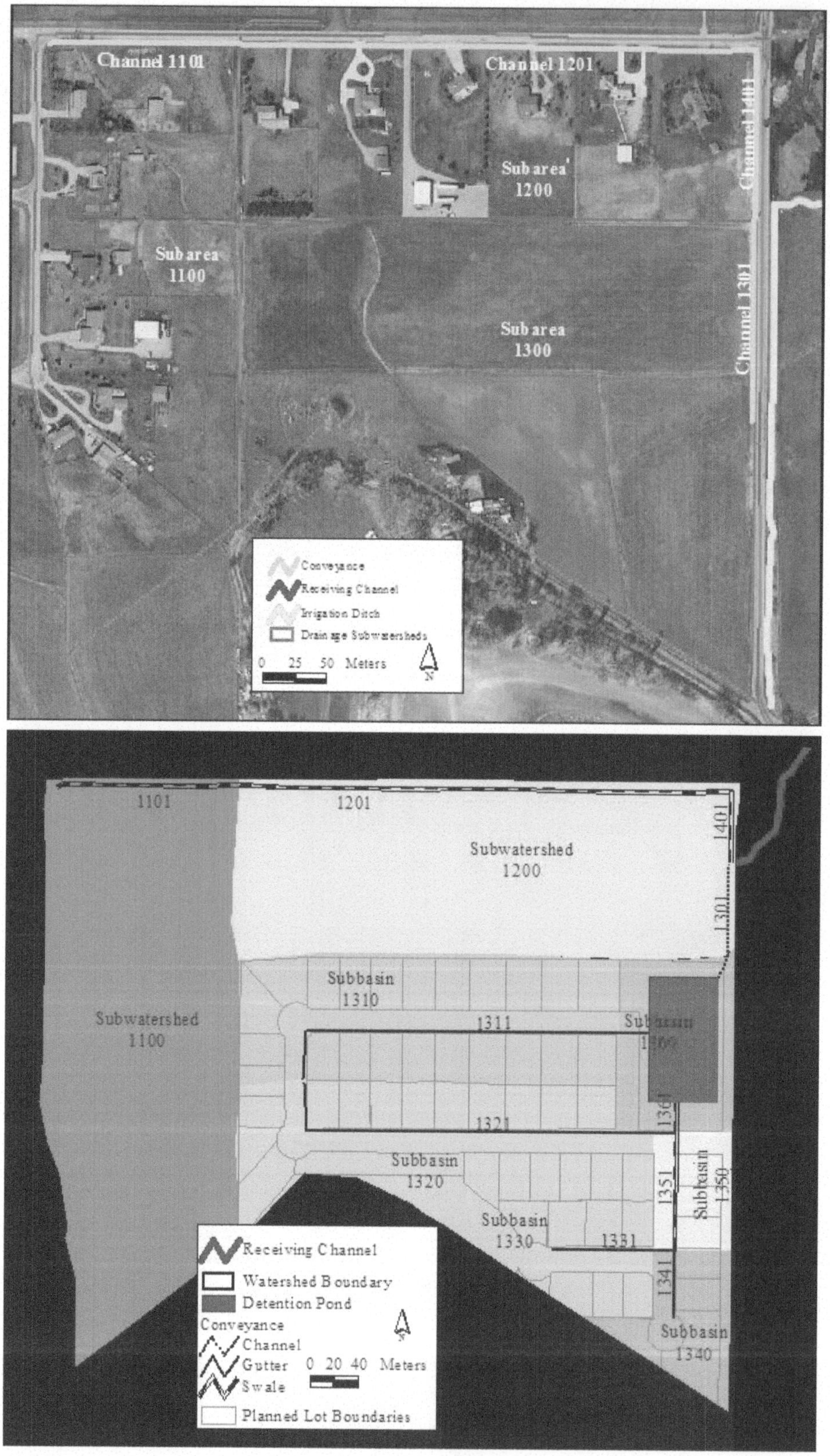

Figure 3-13. Hypothetical Development in Fort Collins, Colorado (a) Existing Conditions (b) Developed Conditions.

Response of Peak Flow Frequency Exceedance and Flow Duration Curves to Alternative Development Scenarios

Figure 3-14 shows the peak flow frequency exceedance curves for five alternative stormwater management scenarios for the hypothetical development in Fort Collins, Colorado. In this figure the Peak Shaving + BMP scenario includes control of the 100- and 10-year storms. The effects of the over control scenarios are evident in the large reduction of peaks that occur infrequently. The geomorphic impact of peak over control is discussed in the next section. In the Peak Shaving + BMP scenario the peak flow frequency exceedance curve is well matched to undeveloped conditions for the 100- through the 2.5-year return interval design storms, but peaks are over-controlled for smaller storms due to the design of the WQCV outlet based on UDFCD (2001) design standards.

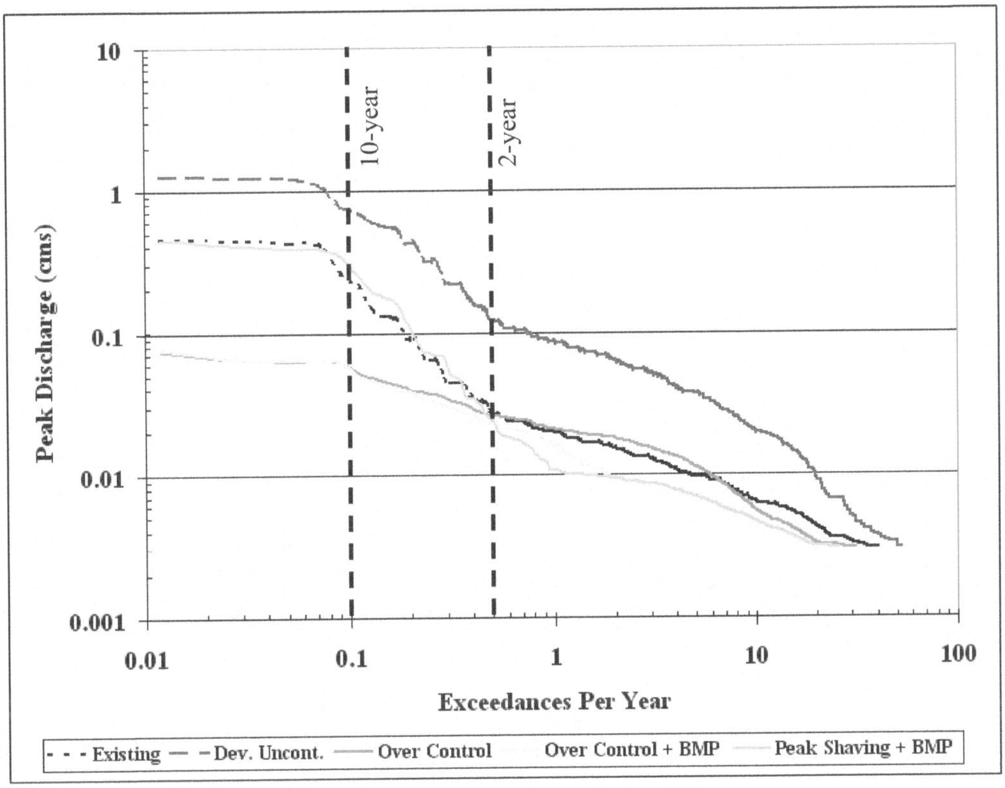

Figure 3-14. Peak Flow Frequency Exceedance in Fort Collins Watershed.

Flow duration curves are also useful tools in evaluating the impacts of stormwater management practices. However, simultaneous post-development reproduction of pre-development peak flow frequency exceedance and flow duration curves is not possible when stormwater detention and retention practices are used. These practices control the magnitude, timing, and duration of flows but do not control the total volume of stormwater runoff. Volumes of stormwater runoff increase when development takes place, and these increased volumes are controlled by changes in the ability of the watershed to infiltrate or evapotranspirate as much water.

Figure 3-15 displays the percent of time discharges are equalled or exceeded over the 50-year simulation for undeveloped conditions and each post-development level of stormwater control examined. In this figure, 1% represents one percent of the 50-year simulation period, or approximately 183 days; 10% represents 1826 days (5 years); 0.1%, 18.3 days, and so forth. Figure 3-15 shows that when development takes place without any type of stormwater control, the discharges are greater in magnitude than "undeveloped" conditions for approximately 5.2 %, or 949 days (2.6 years), of the overall time period examined. The magnitude and duration of discharges in the "Peak Shaving (100) + BMP" scenario are approximately the same as the "Undeveloped" scenario for the largest discharges for 3 hours of the total duration. The duration of discharges in the "Peak Shaving (100 & 10) + BMP" scenario are less than those in the "Undeveloped" scenario between 0.43 cm and 0.26 cm, which is the maximum discharge rate for the 10-year storm. Beyond this point, "Peak Shaving (100 & 10) + BMP" discharges are greater in magnitude than the "Undeveloped" discharges until convergence with the "Undeveloped" discharges for the remaining 16.8% of the time. "Overcontrol" discharges are less than "Undeveloped" discharges for 19 hours.

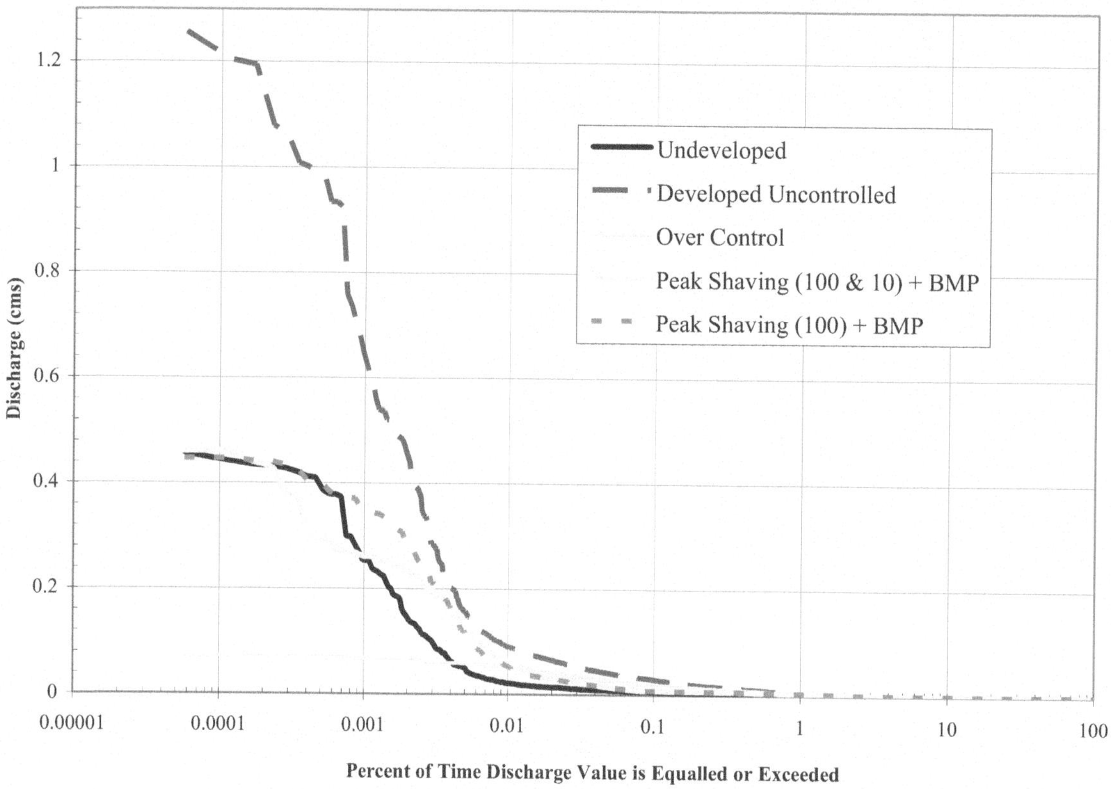

Figure 3-15. Flow Duration Curve for Fort Collins Watershed.

Effect of Alternative Development Scenarios on Erosion Potential for a Variety of Stream Bed Material Sizes

Inability to match the full duration curve does not mean that erosion potential due to changes in flow in the stream has to increase, however, so long as the critical portion of the flow duration curve is matched. The critical portion of the flow duration curve can be described as the duration of flow magnitudes above which stream bed or bank erosion occurs in the channel. A threshold discharge can be calculated through the use of a critical shear stress estimate for the bed and bank materials in the channel. If the duration of flows above the threshold remain the same from pre- to post-development, no increase in erosion potential, as defined in this paper, takes place. It is important to note, however, that changes in sediment supply to the stream channel, such as those that occur due to development, will also affect the potential for erosion in the channel.

The critical portion of the flow duration curve varies depending on the critical shear stress of the stream bed and bank material to be protected. In channels with low sediment entrainment thresholds, e.g., sands, fine gravels, and highly erodible clays with critical shear stresses less than one Pa, the rate of discharge required to entrain particles is so small that the flow duration curve should be controlled in its entirety. For stream bed and bank materials with higher entrainment thresholds, e.g., gravels and resistant clays, the critical discharge value, Q_c, can be estimated by first identifying τ_c, then using the shear stress equation to calculate the hydraulic radius value necessary to initiate particle movement, and, lastly, using the Manning equation to calculate the critical discharge value. The duration of post-development discharges above the critical value should be restricted to the duration of the pre-development discharge. For example, if τ_c for the stream bed materials in the Fort Collins watershed is 7 Pa, the y_c is 0.29 m, Q_c is 0.167 cm and the flow duration curve above this value should be matched to the extent possible.

Using the methods described in Section 3.4.2.4, stream channel erosion potential was evaluated for a variety of critical shear stress values, ranging from 0.2 Pa to 10 Pa, and are presented in Figure 3-16. In Figure 3-16, erosion potential values less than one indicate a decrease in erosion rates pre- to post-development, and erosion potential values greater than one indicate an increase in erosion rates. Erosion potential values equal to one indicate no change in erosion rates. Figure 3-16 demonstrates what was difficult to discern in Figure 3-15: that the use of stormwater detention increases durations of the discharges that entrain sediment with shear stress values less than 0.7 Pa. When τ_c is between 0.2 and 0.7 Pa, the calculated erosion potential is larger when stormwater controls are in place than when stormwater is left uncontrolled. These results indicate that detention could exacerbate erosion in a stream channel above what might occur if the stormwater were left uncontrolled. This increase in erosion potential suggests that a reduction in the total volume of runoff, through the use of infiltration, for example, may provide a better method for stormwater management for watersheds that drain to fine-grained noncohesive or erodible clay streams.

Figure 3-16 also shows that as the τ_c values increase, stormwater management practices create erosion potentials closer to the ideal value of 1.0. The erosion potential values of zero for the "Overcontrol" scenario demonstrate that no erosion is predicted under this scenario if the τ_c is greater than 4. Using the example of τ_c equal to seven Pa, Figure 3-16 shows that the "Peak Shaving (100+10) + BMP" scenario results in an erosion potential value closest to 1, indicating that this type of stormwater control may be most appropriate to minimize changes in stream erosion potential when urbanization of the watershed takes place.

Erosion potential in the study reach was also evaluated for a grain size distribution consisting of $d_{16} = 0.505$ mm (coarse sand), $d_{50} = 3.29$ mm (very fine gravel), and $d_{84} = 12.3$ mm (medium gravel). Erosion potential was calculated for individual grain sizes and for the composite grain size distribution. For the composite grain size distribution analysis, average erosion rates for each time step were calculated using the d_{16}, d_{50}, and d_{84} values as well as estimates of the d_{33} and d_{57} values obtained through interpolation. Table 3-3 summarizes the erosion potential index calculated for each diameter, as well as for the composite grain size distribution. The erosion potential values for each diameter correspond to those for similar τ_c

values presented in Figure 3-16. This table shows that as τ_c values increase, the erosion potential increases when stormwater is left uncontrolled. Conversely, as τ_c values increase, erosion potential decreases when stormwater controls are implemented. When the grain size distribution is considered as a whole, the erosion potential is smallest when stormwater is left uncontrolled, similar to the differences between stormwater management scenarios observed for the d_{16} grain size. This is because of the significant contribution the smaller grain sizes make to the overall erosion rate.

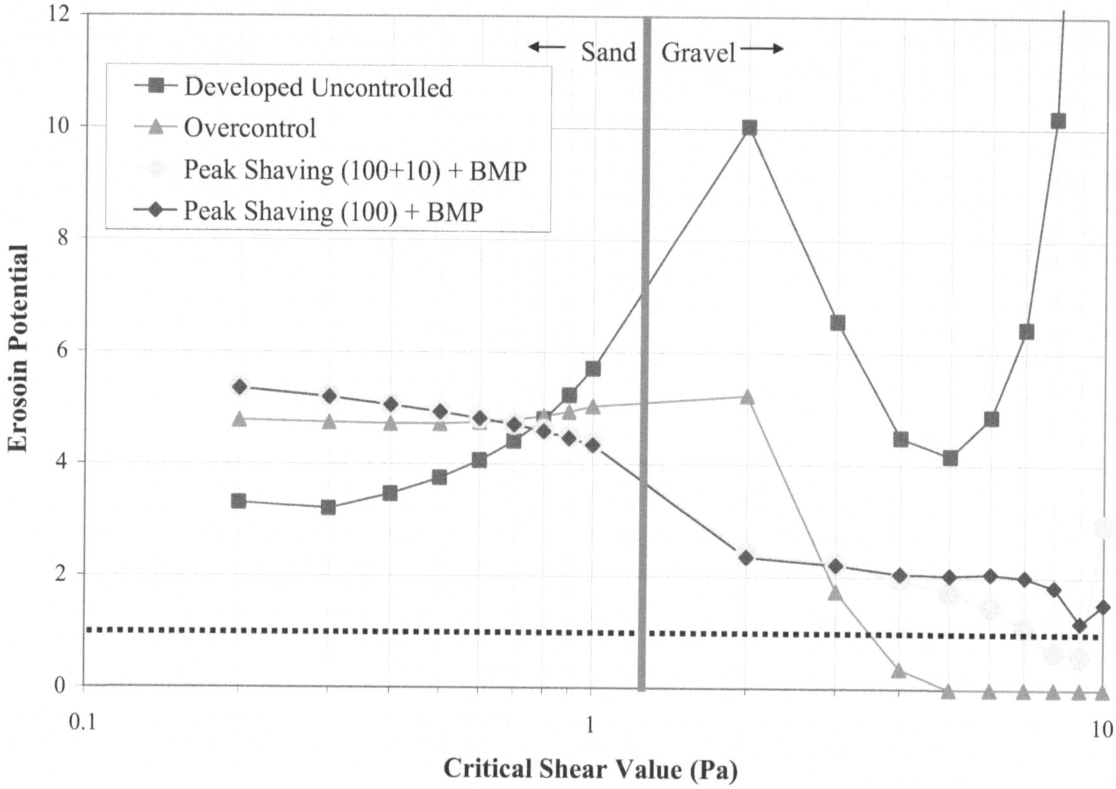

Figure 3-16. Erosion Potential at Varied Levels of Critical Shear Stress. Reprinted from *Water and Science Technology* 54(6-7) p. 352, with permission from IWA Publishing.

Table 3-3. Erosion Potential for Representative Grain Size Distribution. Reprinted from *Water and Science Technology* 54(6-7) p. 352, with permission from IWA Publishing.

Scenario	d_{16} = 0.51 mm τ_c = 0.27 Pa	d_{50} = 3.3 mm τ_c = 2.2 Pa	d_{84} = 12.3 mm τ_c = 9.1 Pa	Full Distribution
Developed Uncontrolled	3.12	9.71	24.54	3.86
Overcontrol	4.75	4.5	0.00	4.80
Peak Shaving (100+10) + BMP	5.24	2.35	0.60	4.98
Peak Shaving (100) + BMP	5.24	2.28	1.14	4.96

These results show that erosion index values are highly sensitive to the critical shear stress values required to initiate erosion. Selection of the stormwater controls best suited to minimize changes in erosion potential requires the identification of these critical shear stress values. Identification of these critical shear stress values requires field evaluation through the use of sieve analysis or direct measurement of erodibility through the use of methods such as jet testing.

Response of $T_{0.5}$ Hydrologic Metric to Alternative Development Scenarios

Figure 3-17 shows the response of the *modified* $T_{0.5}$ hydrologic metric to urban development with varying levels of controls. In this analysis, the *modified* version of the $T_{0.5}$ metric was computed using a fixed pre-development 0.5-year peak flow rate to standardize flow duration curves. This approach is consistent with that used for metrics computed by Cassin et al. (2005) in the King County Normative Flows Project: the T_{Qmean} and the % of time above 2-year mean flow. As expected, the controls were most effective at mitigating the effects of urban development on *modified* $T_{0.5}$, since they were designed to prevent post-development flow rates from exceeding the target flow rates. The peak shaving + BMP controls also appear to bring this metric closer to pre-development levels than the other designs. This makes sense because BMPs limit the range of possible flows by controlling smaller storms which are closer in size to the half-year storm. Also, by reducing the 100-year peak to the pre-development 2-year peak discharge, the over-control designs increase the durations of intermediate flows by releasing the excess volumes at lower rates for longer durations than the peak shaving designs.

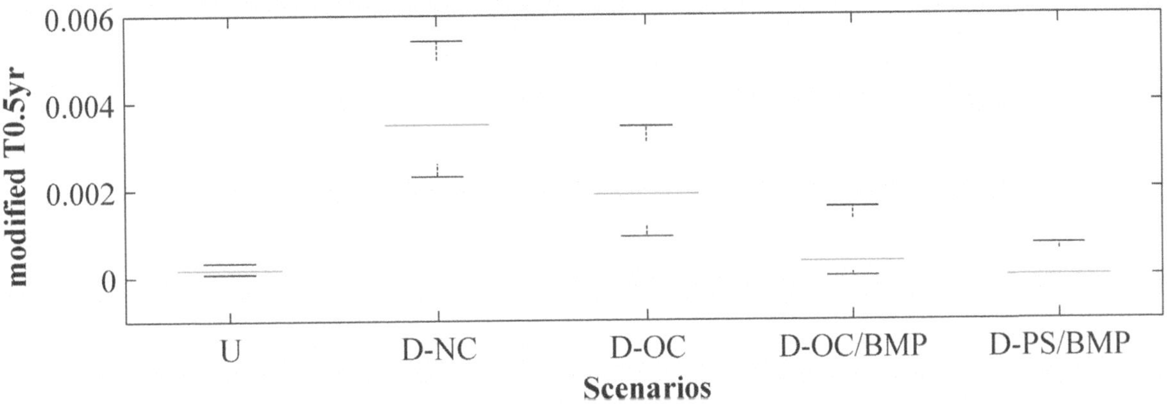

Figure 3-17. Median and interquartile range values of modified $T_{0.5}$ for multiple scenarios of development. U=Undeveloped, D-NC=developed, uncontrolled, D-OC=Overcontrol, D-OC/BMP: Overcontrol + BMP, D-PS/BMP: Peak Shaving (100 & 10) + BMP.

3.6.2.2 Impact of Runoff Control Strategies on Geomorphic Metrics in Lenexa, Kansas

Geomorphic metrics were used in Lenexa, Kansas, to develop design criteria for regional detention basins so that the current geomorphic condition in the natural stream system would be preserved to the maximum extent practicable (Rohrer et. al., 2005). Fifty-three years of historical precipitation data were applied to the EPA SWMM model to conduct continuous hydrologic and hydraulic simulations of a portion of the Cedar Creek watershed in Lenexa, Kansas, generating continuous stream flow hydrographs in the receiving stream channels.

The study area is 93 hectares and currently undeveloped, consisting mostly of pasture land and natural prairie. Development plans for the watershed include the addition of a business park and medium density residential neighborhoods that will increase imperviousness (approximately 2%, currently) to an average of 57 %. The study area was divided into 12 subbasins, as shown in Figure 3-18.

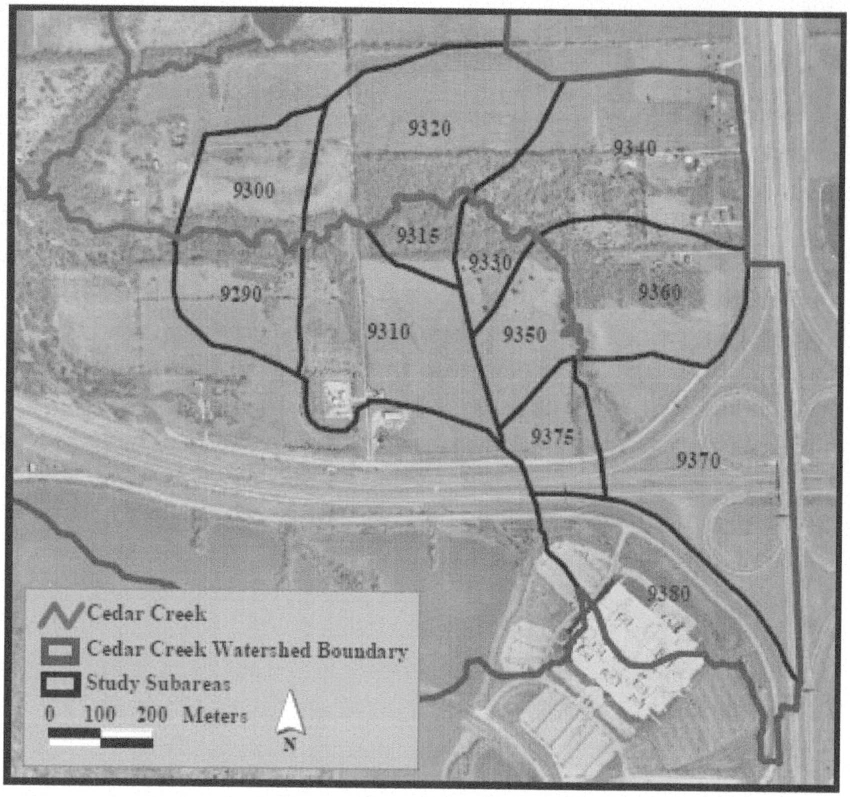

Figure 3-18. Cedar Creek Watershed Study Area.

A number of levels of stormwater detention were examined. Peak shaving was applied for various combinations of the 1-, 2-, and 10-year design storm events. The application of stormwater BMPs was also examined. The term BMP represents control of the WQCV, which was computed as the product of the 90% rainfall depth and a runoff factor based on percentage impervious area using the method described by the Mid-America Regional Council's and American Public Works Association's *Manual for Best Management Practices for Stormwater Quality* (2003). Varied BMP draw-down times were also examined. Results for the following three scenarios are described in the remainder of this section:

- In the "Undeveloped" scenario, all subbasins in the study area are natural, with an estimated imperviousness area of 2%. This scenario assumes the basin is drained by natural channels.
- The "Developed Uncontrolled" scenario is based on future land use conditions (residential and commercial) with a typical stormwater conveyance system consisting of concrete pipes that drain to the natural stream system.
- The "Developed, Controlled" scenario includes peak shaving of the 1- and 10-year storm and control of the WQCV using extended detention with a 40-hour draw-down time.
- The "Developed, Controlled" scenario represents onsite BMP and regional detention criteria that were selected to allow post-development replication of pre-development flow frequency curves and the critical portions of the shear stress duration curves.

Critical shear stress values were identified during field surveys. Table 3-4 summarizes the soil types observed and the literature reference used to identify critical shear stress values for each location based on observations of soil types and channel conditions in the field (Kwiatkowski, 2004).

Table 3-4. Summary of Critical Shear Stress Values and Soil Types in the Cedar Creek Watershed.

τ_c (Pa)	Stream Location	Representative Soil	Reference
12.4	Upper Trib.	colloidal alluvial silts	Chow, 1959
38.3	Main Stem	graded silts to cobbles	U.S. Department of Agriculture, 2001

Instream continuous stage data generated by the SWMM model were used to examine stream channel erosion potential through the use of the erosion potential index described in Section 3.4.2.4. Critical flow thresholds were extrapolated from critical shear stress values. These critical flow thresholds were used to identify return frequencies above which the duration of stormwater must be controlled to pre-development levels.

Similar to a flow duration curve, as discussed in Section 3.6.2.1, it is important to match the critical portion of a shear stress duration curve to minimize changes in erosion potential in a stream channel. The critical portion of the shear stress duration curve can be described as the duration of shear stress magnitudes above which stream bed or bank erosion occurs in the channel. If the duration of flows above the critical shear stress threshold remains the same from pre- to post-development, no increase in erosion potential takes place.

Figure 3-19 displays the percent of time in-channel shear stresses are equaled or exceeded over the 53-year simulation for "undeveloped" conditions, "developed, uncontrolled" conditions, and the recommended "developed, controlled" conditions. Figure 3-19 shows that when development takes place without any type of stormwater control, the "developed, uncontrolled" shear stress values are greater in magnitude than "undeveloped" conditions for approximately 10.2 % of the time, or approximately 5.5 years of the overall time period examined. The magnitude and duration of shear stresses in the "developed, controlled" scenario are greater in magnitude than "undeveloped" conditions for approximately 18.5 % of the time, or approximately 10 years of the overall time period examined.

The τ_c values determined for Cedar Creek are shown on Figure 3-19, and shear stresses above these critical values were those targeted for control. This figure shows that the duration of time the τ_c value for the main stem (38.3 Pa) is equaled or exceeded increases by an order of magnitude between "undeveloped" and "developed, uncontrolled" conditions. In "undeveloped" conditions, the τ_c value for the main stem (38.3 Pa) is equaled or exceeded 0.02% of the time (three days), and in "developed, uncontrolled" conditions, τ_c is equaled or exceeded 0.23% of the time (45 days). The "developed, controlled" condition allows a closer duration match to "undeveloped" conditions with the critical shear stress threshold τ_c equaled or exceeded 0.03 % of the time, or six days. Figure 3-19 also shows that if the τ_c value for the reach were a lower value, such as the value determined for the upper tributaries, 12.4 Pa, the critical portion of the shear stress duration curve would not be as well matched. For this critical threshold, the duration of time τ_c is equaled or exceeded ranges from 0.08% (16 days) in "undeveloped" conditions, 0.54% (105 days) in "developed, controlled" conditions, and 2.10% (394 days) in "developed, uncontrolled" conditions. The order of magnitude increases in the duration of time at this lower critical shear stress value, regardless of the stormwater control, demonstrates the importance of identifying the critical shear stress threshold when designing stormwater management practices

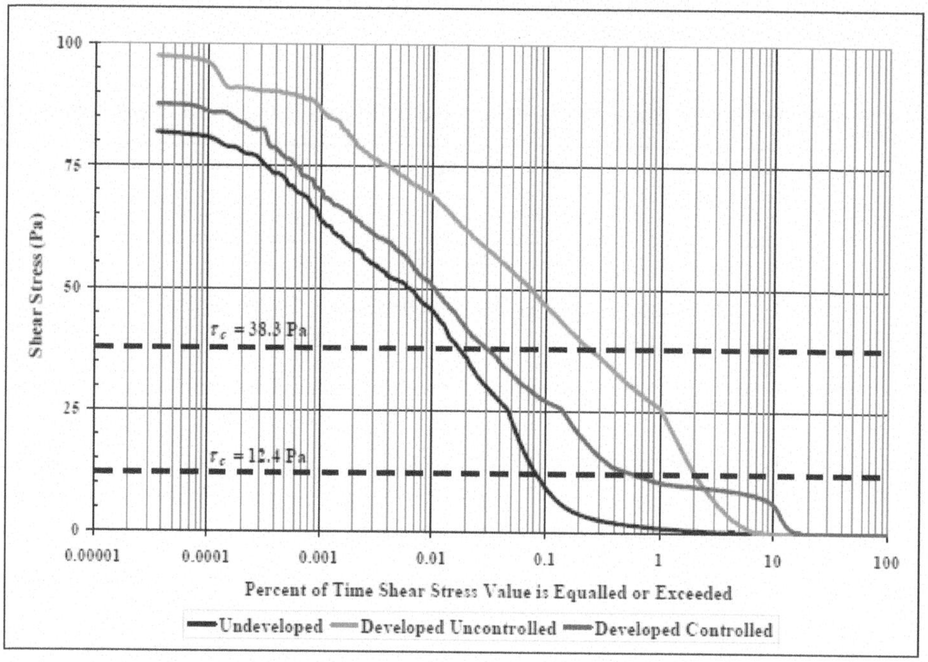

Figure 3-19. Shear Stress Duration in Cedar Creek Watershed.

to minimize stream channel erosion potential.

Table 3-4 quantitatively describes differences in excess shear using the stream channel erosion potential index discussed in Section 3.4.2.4. Table 3-5 shows that for both τ_c values, erosion potential is decreased with the application of the "developed, controlled" scenario that includes peak shaving of the one- and ten-year storm and control of the WQCV using extended detention with a 40 hour draw-down.

Critical flow thresholds were extrapolated from the critical shear stress values determined for the upper tributaries and main stem of Cedar Creek. These critical flow thresholds were used to identify return frequencies above which the duration of stormwater must be controlled to pre-development levels. The return periods for these critical flow thresholds are shown in Figure 3-20. It should be noted that because shear stress is dependent upon channel geometry, the critical discharges presented in Figure 3-20 are for the downstream-most channel in the study area only. Figure 3-20 shows that the flow frequency curve for the "developed, controlled" conditions approximately matches the flow frequency curve for the "undeveloped" conditions as designed for the one- and 10-year return interval design storms. This match of the flow frequency curve, however, does not indicate a match in the duration of critical flows, as can be inferred from the previous evaluation of the shear stress duration curves.

Table 3-5. Stream Channel Erosion Potential under Developed Conditions.

Scenario	Erosion Potential	
	τ_c=12.4 Pa	τ_c=38.3 Pa
Developed, Uncontrolled	19.0	13.0
Developed, Controlled	2.7	1.8

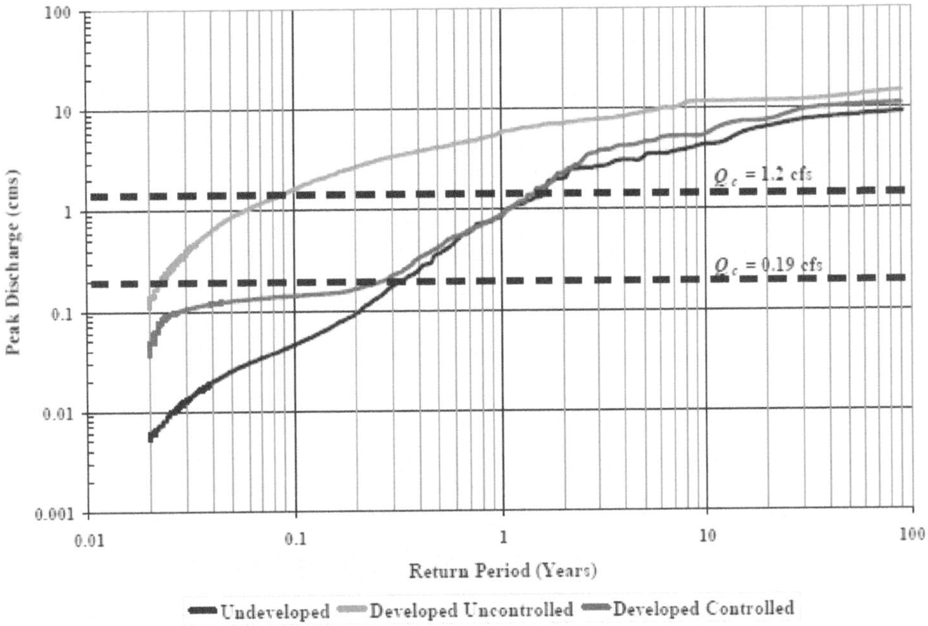

Figure 3-20. Peak Flow Frequency Exceedance Curve for Cedar Creek Watershed.

3.7 Step 7 - Establish Management Criteria for Stream Types of Interest

This step (Figure 3-21) is the administrative process of setting stormwater management criteria and is basically the design criteria for runoff controls on new development. The rules, which will be based on the extensive analyses examined in this document, should be straightforward and understandable by consulting engineering firms that do development work. Criteria can be set by municipal administrators based on analysis of relationships established by the protocol described in the previous sections. Stormwater management practices can be evaluated to identify those which provide hydrologic or geomorphic metric results in the targeted range. An example of this comparison is illustrated in Figure 3-22.

In all the applications investigated by the CSU study team, which include moderate rainfall areas like eastern Kansas and semi-arid areas like Fort Collins, Colorado, it appears that if drainage criteria equivalent to the following are required for developing watersheds that the negative hydrologic and geomorphic impacts due to urbanization will be reduced, albeit not eliminated:

- Control of the peak runoff from the 100- and 10-year post-development flows to the peak flows in the pre-development condition.
- Control of the post-development peak runoff from a small storm with return interval in the range of 1 to 2 years to the peak flow in the predevelopment condition. The correct storm should be determined by the stormwater management agency by fitting the pre-development peak flow-frequency exceedance curve as discussed in this report.
- BMPs that have equivalent hydraulic performance to that of an extended detention basin, sized in accordance with the WEF *Manual of Practice for Urban Stormwater Management* (WEF& ASCE, 1998) and having a 40-hour draw-down time.

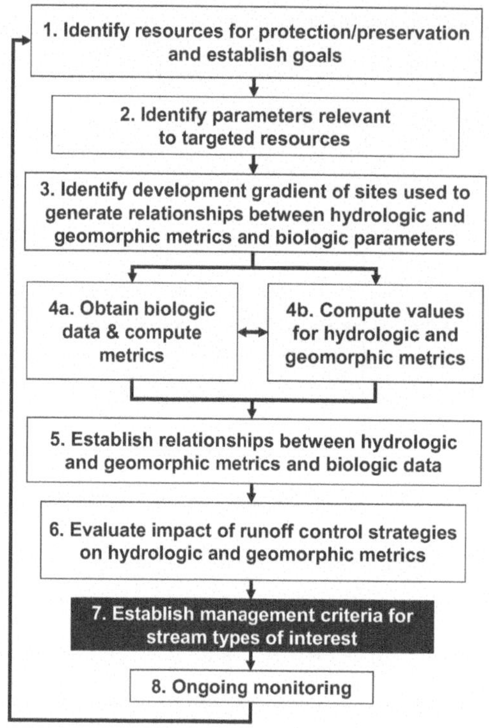

Figure 3-21. Establish Management Criteria.

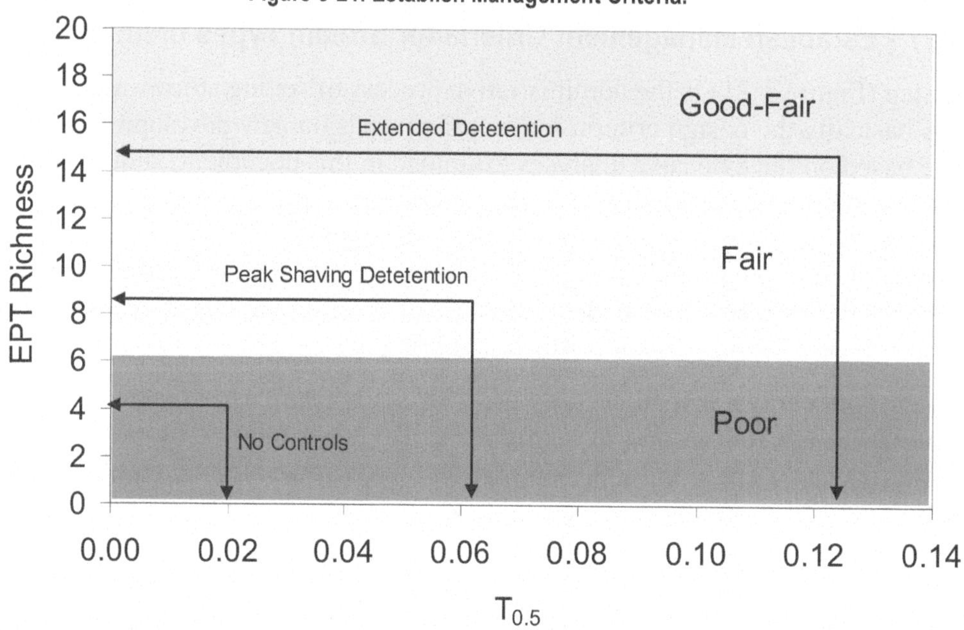

Figure 3-22. Evaluating the Effects of Stormwater Management Criteria on Expected EPT Richness.

It should be noted that volume controls, such as infiltration, were not examined in this study because of their limited applicability. Use of such devices can be evaluated using the protocol and may change some of the above recommendations for runoff control measures in developing watersheds. Low-impact development (LID) techniques were also not evaluated; however, if LID applications have an overall hydraulic effect on catchment runoff that is similar to the extended detention basin prescribed above, then the LID system should perform similarly to the ones tested here.

Finally, many smaller jurisdictions do not have the resources to carry out the protocol as developed in this project. For such entities, it is recommended that the runoff control criteria provided above be used.

3.8 Step 8 — Ongoing Monitoring

Once the management criteria have been established and implemented in Step 7, Step 8 of the protocol calls for ongoing surveillance monitoring to evaluate the effectiveness of protection efforts. Given a biological condition gradient in which to frame further study

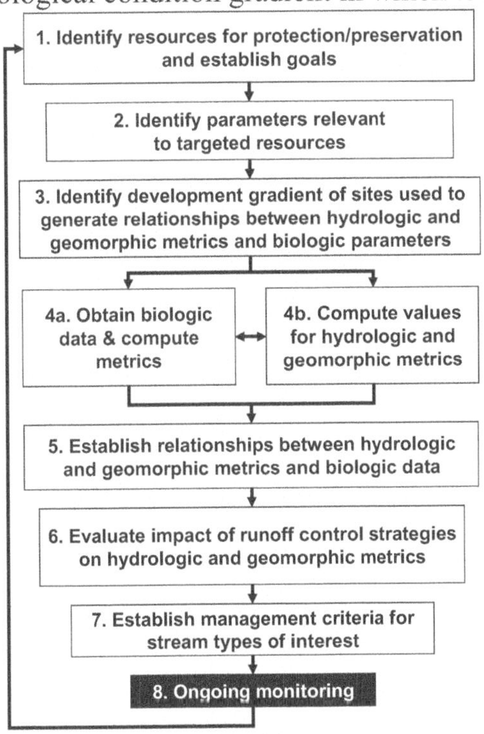

Figure 3-23. Ongoing Monitoring.

observations, surveillance monitoring can focus on the evaluation of watershed(s) targeted for protection or restoration. Surveillance monitoring efforts are intended to evaluate biological condition and the status of hydrologic, habitat, and geomorphic parameters in the protected watershed(s) to determine whether management measures are adequately protective. The sampling protocols identified and discussed in the previous sections should generally be appropriate for ongoing surveillance monitoring, with an increased emphasis on the protected watersheds.

In the event that one or more hydrologic, geomorphic, habitat, or biologic parameters fails to meet established protection goals for the target watersheds, an adaptive management approach is recommended. Sampling parameters and associated protocols should be refined at that point for the purpose of identifying the stressor(s) resulting in nonattainment of biotic goals. In some cases, management criteria derived from the initial protocol implementation could prove ineffective against one or more environmental stressors. In this event, the protocol should be repeated for the refinement of biotic goals (Figure 3-23, Step 1) or refinement of selected measurement parameters and associated linkages (Figure 3-23, Steps 2 through 6). Management criteria might then be revised or replaced with alternative measures to more effectively address the identified stressor(s).

CHAPTER 4.0

PROTOCOL TESTING IN THE NORTH CAROLINA PIEDMONT

4.1 Selection of Pilot Area

The protocol was applied in the Raleigh, North Carolina, area using data gathered by the USGS under the NAWQA program. Specifically, data collected in the North Carolina Piedmont under the study *Effects of Urbanization on Stream Ecosystems* (EUSE) were used for application and testing of the protocol described in this report. The objectives of the USGS EUSE study are to determine the physical, chemical, and biological responses of stream systems to a gradient of increasing urban intensity; determine the most important landscape features driving hydrologic, chemical, geomorphic, habitat, and biological responses to urbanization; determine the physical and chemical factors associated with biological responses; and compare these responses among metropolitan areas located in different geographic settings (climate, geology, hydrology, biology, water use) in 11 regions across the United States (USGS, 2007).

The Raleigh, North Carolina, region was chosen for application and testing of the protocol based on the size of the study watersheds and the willingness of the USGS to share the data at the time it was needed for testing of the protocol. Nearly half of the 30 watersheds investigated by the USGS in the Raleigh, North Carolina region are less than 5 square miles, a size that is practical for modeling at the level of detail needed to produce data useful for calculating hydrologic and geomorphic metrics. This watershed size is also the typical size that a stormwater management agency would consider for application of the protocol. Additional factors which discouraged selection of alternate USGS study regions for application of the protocol included the influence of combined sewer overflows and the influence of transbasin diversions in arid and semiarid regions.

Thirty North Carolina watersheds were screened against the site selection criteria summarized in Section 4.3, including identification of watersheds with homogeneous land use (relatively uniform, rather than mixed land use within the watershed), streams with similar geomorphic characteristics, watershed size, and the elimination of watersheds with confounding factors such as point source discharges and dams. Watersheds were also selected based on the degree of urbanization in each basin as represented by multi-metric urban intensity index (UII) values computed by the USGS (McMahon & Cuffney, 2000; Coles et al., 2004; Tate et al., 2005). The UII includes census, socioeconomic, infrastructure, and land-use and land-cover metrics that correlate with population density. The variables evaluated were population density, household density, road density, percentage of basin in urban land, and percentage of buffer (90 m on either side of the stream) in urban land. These characteristics were range standardized over all sites, and characteristics that were negatively correlated with population density were reversed. At each site, the five range-standardized values were averaged to produce a ranking of relative urbanization at each site. Finally, the resulting rankings were range standardized over all sites and multiplied by 100 to produce an index that varies from 0 to 100, with 0 representing minimal urbanization and 100 representing maximum urbanization (Giddings & Steuer, 2007).

Ten watersheds were initially selected for application of the protocol; these watersheds range in size from 3 to 9 square miles. The selection of several watersheds larger than the recommended size criterion of 5 square miles illustrates the nature of the above criteria as a recommended point of departure and not absolute rules for site selection. Ultimately, two of the 10 selected watersheds were eliminated from the protocol application. One was eliminated based on the recommendation of USGS NAWQA staff in North Carolina because of heavy development that took place within the watershed during the time of the EUSE study. The other watershed was eliminated due to time constraints of CSU project team staff during field surveys. Table 4-1 lists the selected NAWQA study watersheds, their corresponding drainage area size, and their associated level of urbanization/development as represented by the UII. The locations of these watersheds are shown in Figure 4-1.

Table 4-1. Selection Characteristics of Watersheds Chosen for Protocol Application.

Watershed	USGS Site Number	Basin Size (sq mi.)	Urban Index (UII)	Total Impervious Area (%)	Homo-geneity (0 = Low 1 = Med 2 = High)	Stream bed Material % Silt / Clay / Sand	% Gravel / Cobble
Morgan Creek	02097464	8.3	4.9	0.8	2	16.3	58.2
Strouds Creek	0208501535	8.9	7.2	3.8	1	3.6	81.9
Camp Branch	0208794025	3.2	12.9	2.9	1	16.4	78.1
Wilson Creek	0209750881	3.5	24.1	9.0	1	34.6	63.6
Dutchman's Branch	0208758440	4.6	28.6	4.4	1	9.1	90.9
Black Creek	0208725055	3.6	79.6	20.5	2	23.6	16.3
Beaverdam Creek	02087304	3.0	90.8	11.7	2	40.0	58.2
Pigeon House Creek	0208732610	4.5	100.0	30.6	2	30.9	58.2

4.2 Data Development and Watershed Model Calibration

4.2.1 Data Collection

Data measured by the USGS at the EUSE sites included physical stream characteristics such as physical habitat, stream stage, and water temperature; chemical constituents, including nutrients, pesticides, organic carbon, suspended sediment, sulfate, chloride, dissolved oxygen, pH, specific conductance in water, organic contaminants, and potential toxicity; and biological integrity including invertebrate, algae, and fish communities, and algal biomass (USGS, 2007).

Benthic macroinvertebrate data were collected by the USGS in the North Carolina pilot area from early May to late June 2003 using standard NAWQA program sampling protocols (Moulton et al., 2002). Macroinvertebrates were collected using two types of samples — a quantitative sample collected from instream riffles (richest targeted habitat, RTH) and a qualitative multihabitat (QMH) sample that collected invertebrates from as many habitats in the stream reach as were accessible (Giddings & Steuer, 2007). Figure 4-2 shows EPT richness values for the RTH for the 30 watersheds in the North Carolina Piedmont. Ratings for various EPT richness scores are given in Table 4-2. It should be noted that only one of 30 samples generated an EPT richness rating of good–fair with the remainder ranking as fair or poor. This lack of a full range of biotic integrity is perhaps due to historic land use within the North

Carolina Piedmont; while a number of the undeveloped watersheds in the study area were forested or partially forested at the time of the study, it is likely that these watersheds were historically used for agriculture.

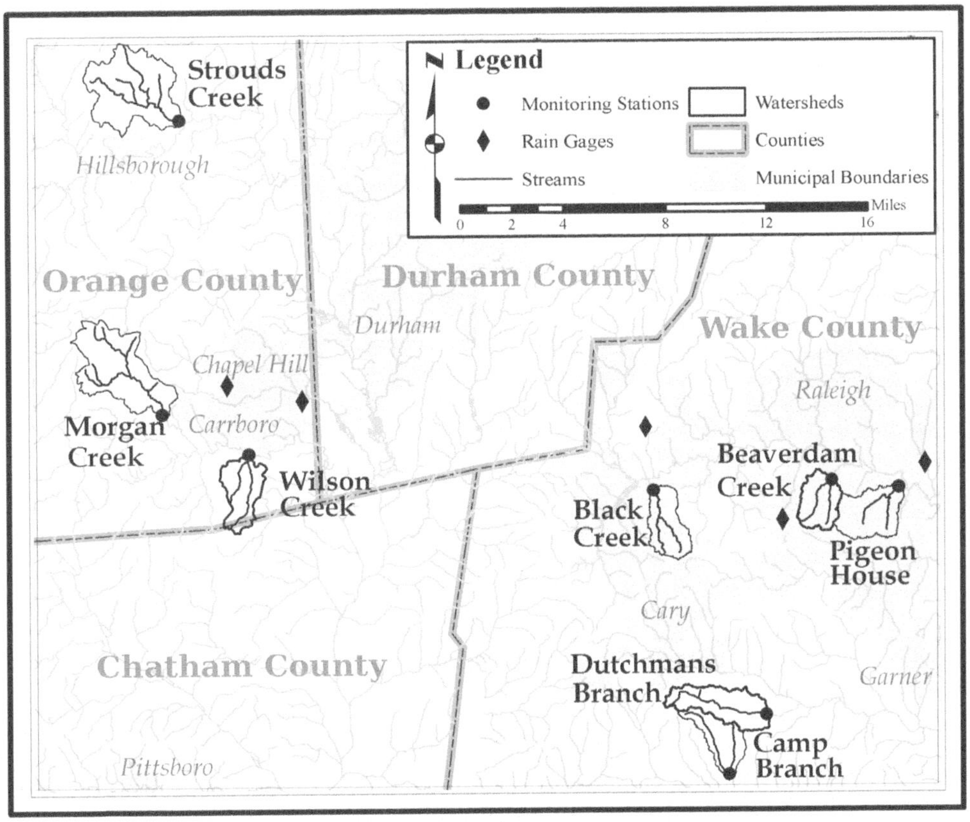

Figure 4-1. Pilot Study Watersheds.

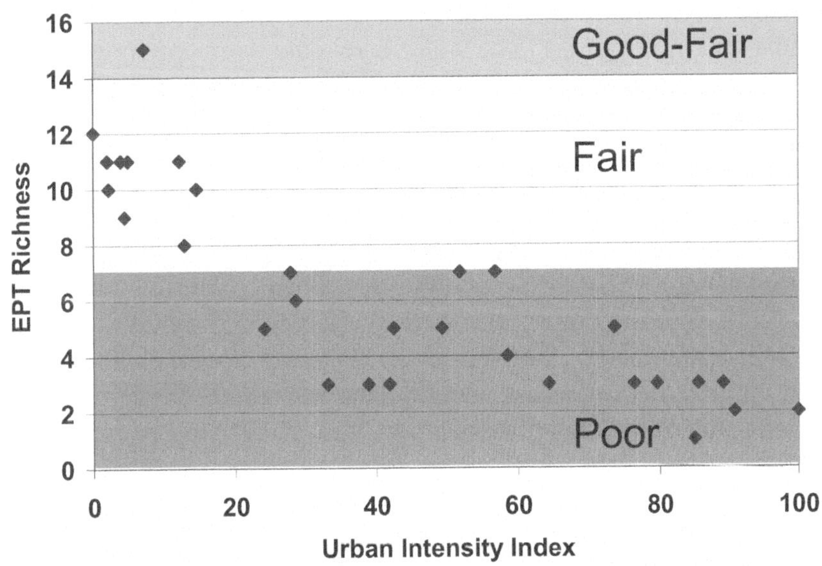

Figure 4-2. Range of EPT Richness Scores in North Carolina Pilot Area.

Table 4-2. EPT Richness Rating Scores (NCDENR, 2006).

Rating	EPT Richness
Excellent	>27
Good	21-27
Good-Fair	14-20
Fair	7-13
Poor	0-6

Bed substrate composition was evaluated by the USGS in the pilot watersheds using methods described by Fitzpatrick et al. (1998). Characteristics at the reach scale were measured at 11 equally spaced transects along the sampling reaches (USGS, 2007) and five samples were evaluated at each transect. Table 4-3 summarizes the bed substrate data collected by the USGS and shows the percent of samples with the dominant substrate in each size class for each watershed.

Table 4-3. Summary of Dominant Bed Substrate in the Pilot Watersheds.

Class Name	d_s (mm)	Strouds	Morgan	Camp	Wilson	Dutch-mans	Black	SW Prong	Pigeon
Silt/clay/marl/muck/ organic detritus	-	4%	13%	9%	7%	6%	15%	6%	0%
Sand	>0.063	0%	4%	7%	27%	4%	9%	35%	31%
Fine/medium gravel	>2	18%	4%	13%	15%	36%	2%	22%	22%
Coarse gravel	>16	6%	6%	35%	26%	22%	4%	18%	20%
Very coarse gravel	>32	11%	22%	20%	20%	16%	2%	16%	6%
Small cobble	>64	22%	22%	9%	4%	15%	4%	0%	7%
Large cobble	>128	26%	6%	2%	0%	2%	6%	2%	4%
Small boulder	>256	11%	7%	2%	0%	0%	7%	0%	2%
Other	-	4%[1]	18%[1]	4%[2]	2%[1]	0%	53%[1]	2%[1]	9%[1]

[1]Large boulder, irregular bedrock, irregular hardpan, or irregular artificial surface, d_s > 512 mm
[2]Smooth bedrock, concrete, or hardpan.

Hydrologic data were collected by the USGS at the outlet of each of the EUSE watersheds using pressure transducers. Water level data were collected at 5-minute increments for a portion of the monitoring period and at 15-minute increments for the remainder of the monitoring period. For the eight watersheds applied in the protocol, the period of record varied from 12 months to 17 months depending on the monitoring station. Three of the eight stations contained data gaps of more than 1 day, and these are indicated in Table 4-4. The period of record for each station is summarized in Table 4-4. The water level data were converted to discharge values by USGS staff using stage-discharge rating curves established using HEC-RAS models (Giddings & Steuer, 2007). Computed discharge values were not provided by the USGS for Strouds Creek due to problems with measured cross-sectional data and the HEC-RAS model for that watershed.

Field data were also collected by the CSU project team. In November of 2005, stream cross-section data were measured using level and rod methods described by Harrelson et al. (1994). Locations of the cross-sections in the eight watersheds were recorded by GPS. Stream cross-sections were measured at approximately 2500-foot intervals along stream channels within each of the watersheds. In addition to collecting cross-section data, in-channel Manning roughness values were identified using guidelines provided by Chow (1959) and Arcement and Schneider (1989).

Precipitation data were obtained for application in the SWMM models. Locations of the rain gages considered for use are shown in Figure 4-1.

Table 4-4. Dates of Water Level Monitoring Data.

Watershed	USGS Site Number	Length of Record (months)	Range of 5-Minute Data		Range of 15-Minute Data		Data Gaps (days)
Morgan Creek	2097464	17.6	N.A	N.A.	15JUN02	30NOV03	-
Strouds Creek*	208501535	15.2	12JUL02	16DEC02	16DEC02	19OCT03	20
Camp Branch	208794025	15.7	30JUL02	18DEC02	18DEC02	19NOV03	-
Wilson Creek	209750881	15.6	31JUL02	16DEC02	16DEC02	19NOV03	-
Dutchman's Branch	208758440	15.7	30JUL02	18DEC02	18DEC02	19NOV03	-
Black Creek	208725055	16.7	01JUL02	16DEC02	16DEC02	21NOV03	-
Beaverdam Creek	2087304	12.6	27JUN02	17DEC02	17DEC02	05AUG03	20
Pigeon House Creek	208732610	15.1	01JUL02	17DEC02	17DEC02	19NOV03	47

Five rain gages were identified in the proximity of the study watersheds with long-term (more than 5 years' worth) data. Of these gages, two had 15-minute data and the others had hourly rainfall data. Two of the gages did not have a desired record length of 20 years, and one other had large gaps in its record. Annual precipitation summaries for the two gages with adequate rainfall records are shown in Table 4-5. One gage, monitored by the National Weather Service (NWS) at the Raleigh Durham International Airport (Coop ID 317069), had hourly rainfall records; the other monitored by the USGS at Marsh Creek near New Hope, NC (Site No 0208732885), had 15-minute rainfall totals. As can be seen, the annual totals for each year are similar, and the overall averages of the totals were very similar.

The USGS record at Marsh Creek near New Hope, NC (Site No 0208732885), contains 13 data gaps of 30 days or longer. Data from NWS gage at the Raleigh Durham International Airport (Coop ID 317069) were examined for the periods of record during these data gaps to determine if it was likely that these data gaps were dry periods or if the USGS gage was not functioning properly during the data gap time periods. Based on this analysis, 11 time periods ranging from 31 to 643 days were identified and filled in with data from the NWS gage. The hourly data from the NWS gage were translated to 15-minute data using NetStorm (Heineman, 2007) and combined with the USGS gage data to form a complete rainfall record. Additionally, the precipitation data from the USGS gage were evaluated to identify those values that seemed out of the normal range of expected values. All records of 1 inch or more were individually examined, and several were modified based on values from the NWS gage or deleted altogether.

4.2.2 Model Development

Watersheds were modeled in EPA SWMM5 using both geographic information system (GIS) and field-determined data. Each watershed was discretized into subcatchments by applying the ArcHydro extension of ArcGIS to 10-m digital elevation model (DEM) data from the USGS Seamless Web site (USGS, 2006a.) Hydrologic characteristics for subareas in each watershed were estimated by area weighting measurements of slope, runoff length, soil type and imperviousness. This section defines the model parameters that were used in SWMM5 and identifies the basic calibration technique used to determine their values; details of the calibration algorithm used in this study are provided in Section 4.2.3.

The first step in model development was to define the stream system to be modeled for each watershed, then divide the watershed into smaller catchments (subareas) that drain to the stream system. Criteria for defining a catchment were based on guidance for SWMM application by Huber and Dickinson (1998) and included uniform catchment slope, surface soil type, and

Table 4-5. Annual Precipitation Summaries.

Year	National Weather Service Raleigh Durham International Airport Coop ID 317069 (inches)	USGS Marsh Creek near New Hope, NC Site No: 0208732885 (inches)
1986	36.95	42.37
1987	42.09	56.03*
1988	37.66	52.44
1989	54.21	55.74*
1990	37.55	39.2*
1991	35.46	36.16*
1992	43.18	43.17*
1993	38.05	39.29
1994	36.41	35.89*
1995	48.59	56.84*
1996	45.02	41.7*
1997	40.84	27.81*
1998	50.14	44.89
1999	50.00	57.26
2000	37.07	39.53*
2001	33.77	28.56
2002	45.98	42.43
2003	47.45	45.13
2004	46.00	33.50
2005	37.53	35.26
Mean	42.20	43.02

*Indicates that the record for that year was revised or supplemented with data from the National Weather Service gage at Raleigh Durham International Airport Coop ID 317069.

homogeneous land use. As a rule of thumb, catchment areas in *undeveloped* basins were targeted at 75 acres, but the size of individual catchments ranged from 30 to 300 acres. The target size for catchment areas in *developed* watersheds was also 75 acres, but catchment sizes varied from 50 to 300 acres. Overland flow lengths in undeveloped catchments were computed in the traditional way by area weighting typical overland flow lengths measured within the catchment. Depending on the catchment geometry, catchments drained either directly to a channel in the main stream system or to a minor drainage way tributary to the main stream system.

For urbanized catchments, the overland flow length was taken as the average length from the back of a residential lot to the middle of the street serving the residence, which results in a very wide catchment. To account for gutter and pipe routing within the catchment, the watershed was drained to a routing channel whose length was determined by following the drainage system from its outfall to the centroid of the catchment and whose cross-section was taken as the cross-section of the catchment outfall, which could be either a pipe or an open channel.

Infiltration was simulated using the Horton infiltration equation. The Horton model takes four input parameters: the maximum infiltration capacity of the soil, the minimum or final infiltration capacity of the soil, a decay coefficient which controls how quickly the soil loses its capacity, and the time that has elapsed from the beginning of the storm. Although the Horton model is the most widely known model for infiltration, it is dependent on discrete field measurements or limited parameters from the literature. In this study, initial estimates of the Horton infiltration values for maximum and final infiltrations were assigned based on SCS (Soil Conservation Society) hydrologic soil groups. Minimum infiltration values (f_c) were estimated by hydrologic soil groups (Huber & Dickenson, 1998) and maximum infiltration values (f_o) were linked to watershed soils through soils textures (sand, loam, silt, clay) described in the SWMM5 model help menu (U.S. EPA, 2006). These infiltration values are listed in Table 4-6. Final values used were determined by calibration of model simulations to observed in-stream flow-duration curves and peak-flow frequency curves in each watershed (see Section 4.2.3 below) and are given in Table 4-7. Infiltration rate decay constants (K) and times for a fully saturated soil to dry completely (T_w) were also based upon U.S. EPA (2006) recommendations from the SWMM5 model help menu and are shown in Table 4-7.

Table 4-6. Horton Infiltration Values.

Hydrologic Group	f_c Ranges (in./hr) High	f_c Ranges (in./hr) Low	Average f_c (in./hr)	f_o (in./hr)
A	0.45	0.30	0.375	7.5
B	0.30	0.15	0.225	4.5
C	0.15	0.05	0.100	3.0
D	0.05	0.00	0.025	1.5

Table 4-7. Watershed Parameters Used in SWMM Models.

Watershed	Imperviousness				Horton Infiltration				Depression Storage		Mannings Overland Roughenss	
	GIA	DCIA	K_1	K_2	f_o (in/h)	f_c (in/h)	T_w (days)	K (1/h)	Imperv.	Perv.	Imperv.	Perv.
Morgan	N.A.	0.8	N.A.	N.A.	4.50	0.40	7.0	4.00	0.05	0.20	0.010	0.30
Strouds	N.A.	3.8	N.A.	N.A.	4.50	0.40	7.0	4.00	0.05	0.20	0.010	0.30
Camp	N.A.	2.9	N.A.	N.A.	4.50	0.50	7.0	4.00	0.07	0.25	0.012	0.30
Wilson	9.0	9.0	1.000	1.000	4.47	0.45	7.0	4.14	0.07	0.26	0.012	0.26
Dutchmans	N.A.	4.4	N.A.	N.A.	4.50	0.60	7.0	4.00	0.07	0.25	0.012	0.30
Black	20.5	8.2	0.136	1.350	4.42	0.60	7.0	2.00	0.07	0.25	0.012	0.30
Beaverdam	11.7	4.9	0.145	1.405	4.21	0.70	7.0	1.50	0.07	0.15	0.012	0.26
Pigeon	30.6	14.3	0.130	1.350	4.23	0.70	7.0	1.50	0.07	0.25	0.012	0.30

N.A. = Not applicable because DCIA was calculated directly from road areas.

The level of imperviousness, or gross imperious area (GIA), of each subarea in the highly urbanized watersheds (Black, Pigeon, and Beaverdam) was computed from the 2001 National Land Cover Database Zone 60 Imperviousness Layer (USGS, 2003). In this study, directly connected impervious area (DCIA) was modeled, rather than GIA, in order to more accurately represent the effect of runoff from roofs and parking lots that are routed to pervious surfaces. The DCIA for the urban catchments was estimated using the following equation (Lee and Heaney, 2003).

$$DCIA = K_1 \cdot GIA^{K_2} \qquad (4.2.1\text{-}1)$$

The original parameters of K_1 and K_2 proposed by Lee and Heaney (2003), 0.15 and 1.41, respectively, were modified for the North Carolina pilot area during the calibration process by comparing simulated peak-flow frequency exceedance curves to peak-flow frequency exceedance curves computed from the level-flow monitoring data collected by the USGS from 2002–2003. The methodology used to develop the peak flow frequency exceedance curves is described in Section 4.2.3. In the pilot watersheds the K_1 and K_2 values ranged between 0.13 and 0.145 for K_1 and 1.35 and 1.405 for K_2. The DCIA in the rural watersheds (Morgan, Strouds, Camp, Dutchman's, and Wilson) was estimated from the area of the roads in each subcatchment as measured from aerial photos (USGS, 2006b). DCIA and GIA values, when calculated, as well as K_1 and K_2 values are given in Table 4-7.

Stream channels were modeled as irregularly-shaped cross sections using data collected by CSU staff during the November 2005 field surveys described in Section 4.2.1. Floodplains were estimated using 2-foot contour data (Wake County, 1999; NCDOT GIS, 2005) to allow for the simulation of overbank flow in the model without creating a "flooding" condition within the model. The bottom of each stream channel segment modeled was also measured from the 2-foot contour maps and channel lengths were calculated using ArcMap GIS.

Fifteen-minute precipitation data from USGS gage number USGS Site No 0208732885 (Marsh Creek near New Hope, NC) were applied to each of the models. A twenty year period of record was modeled from January 1, 1985 to December 31, 2004.

Subsurface flows were found to be important for matching the portion of the flow-duration curve representing the interflow recession limb of runoff hydrographs. Interflow was accounted for in the models by applying the groundwater simulation algorithms in SWMM5 to each of the watersheds. In most watersheds, a single aquifer was modeled for each watershed, with the aquifer accepting surface water infiltration and diverting it to one receiving node near the outlet of the watershed. The Morgan Creek watershed was modeled with three aquifers, with headwater watersheds contributing surface water to receiving nodes near the outlets of the respective tributaries, and the third contributing infiltration from the rest of the watershed to a receiving node near the outlet of the watershed. Groundwater parameters were used to calibrate the models to the recession limb flow conditions. A summary of groundwater parameters used for each watershed is given in Table 4-8 and groundwater equation variables are given in Table 4-9. In general, aquifers were modeled to be comprised of Sandy Clay and parameters from Rossman (2005) were applied. Hydraulic conductivity and the wilting point were used as calibration parameters to allow for a better match of the full flow duration curve, specifically aiding in the match of the recession limb of storm hydrographs. Hydraulic conductivity values greater than one are indicative of a high level of groundwater outflow that would be expected from a loamy sand or sandy aquifer.

Table 4-8. Groundwater Parameters Used in SWMM Models.

Watershed	Porosity	Wilting Point	Field Capacity	Hydraulic Conductivity	Lower GW Loss Rate	Unsat. Zone Moisture
Morgan Creek	0.430	0.100	0.321	2.00	0.000	0.3
Strouds Creek	0.430	0.100	0.321	2.00	0.000	0.3
Camp Branch	0.453	0.085	0.321	3.00	0.000	0.3
Wilson Creek	0.430	0.100	0.321	5.00	0.000	0.3
Dutchmans Branch	0.430	0.221	0.321	0.02	0.002	0.3
Black Creek	0.430	0.221	0.321	0.02	0.0026	0.3
Beaverdam Creek	0.430	0.221	0.321	0.02	0.002	0.3
Pigeon House Creek	0.430	0.221	0.321	0.02	0.002	0.3

Table 4-9. Groundwater Equation Variables Used in SWMM Models.

Watershed	GW Flow Coeff.	GW Flow Exp.	SW Coeff.	SW Exp.
Morgan Creek	0.001	5	0.001	0
Strouds Creek	0.001	5	0.001	0
Camp Branch	0.010	5	0.000	0
Wilson Creek	0.010	7	0.001	1
Dutchmans Branch	0.010	1	0.0001	1
Black Creek	0.010	1	0.001	1
Beaverdam Creek	0.010	1	0.001	1
Pigeon House Creek	0.015	1	0.010	1

4.2.3 Model Calibration

Calibration of seven of the eight watersheds was conducted by comparing model output to the level-flow monitoring data collected by the USGS from 2002-2003. The model for the Strouds Creek watershed was not calibrated because flow data were not provided by the USGS for this watershed. Calibration techniques differed depending on if the watersheds were urban or rural. Urban watersheds (Pigeon, Black, and Beaverdam) were calibrated mainly with adjustments in DCIA, while the rural watersheds (Morgan, Camp, Dutchman's, and Wilson) watersheds were calibrated with adjustments in infiltration. As stated in the previous section, groundwater parameters were used to calibrate the models in lower flow conditions that occurred between rainfall events.

Traditionally, stormwater model calibration is preformed on six or more individual storms of varying size (Akan & Houghtalen, 2003); it is considered complete for the single storm event when the peak discharge and volume of runoff are accurately reflected by the model. Because of the importance of accuracy of the full spectrum of flows in computation of hydrologic and geomorphic metrics necessary to create linkages between land use patterns and runoff control practices to stream biologic integrity, models were calibrated across the full spectrum of flows that occurred during the 2002–2003 period of level-flow monitoring by the USGS to the extent possible.

Model parameters were adjusted to match flow-duration and peak-flow frequency exceedance curves for 15-minute model output to 15-minute level-flow data measured by the USGS at the watershed outlets. Any 5-minute data collected by the USGS during the 2002–2003 level-flow monitoring were converted to 15-minute data for the comparison. The flow duration curves were the primary tool for calibration in the mid- to low-flow range and peak flow frequency exceedance curves were used for calibration of the magnitude of runoff event peaks. The flow-duration curve is a graph of all the discharges during a continuous record and their cumulative exceedances, or the percent of time each discharge occurs during the period of record. Figure 4-3 shows the flow duration results from the calibration of the Morgan watershed. Flow duration curves for the period of calibration for the six remaining watersheds are given in the Appendix.

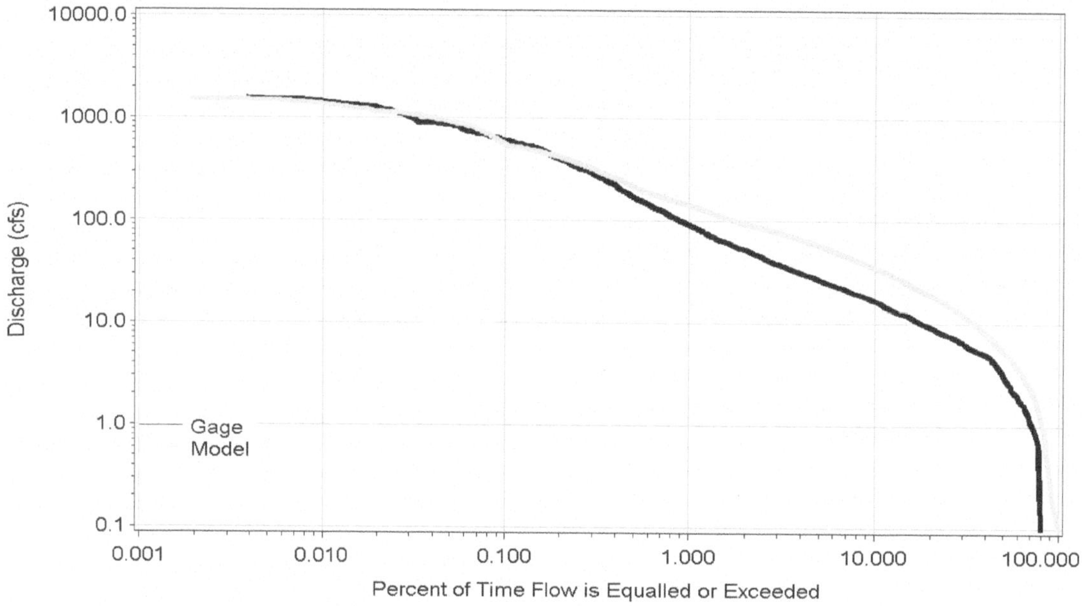

Figure 4-3. Morgan Creek Watershed Flow Duration Calibration Period.

Peak flow frequency exceedance curves were developed from the partial duration series of peak flows. This approach, which is in contrast to the examination of the annual maximum series, was used because it allowed for the analysis of high frequency, low runoff producing storms. In addition, this approach describes the frequency with which a given flow probably occurs as opposed to simply describing the annual maximum flow. A 6-hour inter-event time was specified to separate the flow data into individual events. A sensitivity analysis was performed on flow data from each watershed to identify the minimum threshold to define an event. These thresholds varied from 0.9 to 20 cubic feet per second. Figure 4-4 shows the peak flow frequency exceedance results from the calibration of the Morgan Creek watershed. Peak flow frequency exceedance curves for the period of calibration for the six remaining watersheds are given in the Appendix.

The Cunnane (1978) formula was used to calculate the frequency of an event peak flow, T, as:

$$T = \frac{N+1-2A}{M-A} \tag{4.2.2-1}$$

where T = return interval (years), N = number of years of record, M = rank of the event (in descending order of magnitude), and A = plotting position parameter (0.4). The return interval was converted to exceedances per year, E, using:

$$E = \frac{1}{T}. \tag{4.2.2-2}$$

4.2.4 Development of Hydrologic and Geomorphic Metrics

Twenty years of precipitation data were applied to the eight SWMM models to generate continuous flow hydrographs at the outlets of the pilot watersheds. These continuous flow records were used to compute hydrologic and geomorphic metrics for each of the watersheds.

The $T_{0.5}$ hydrologic metric was computed using SAS from discharges generated by

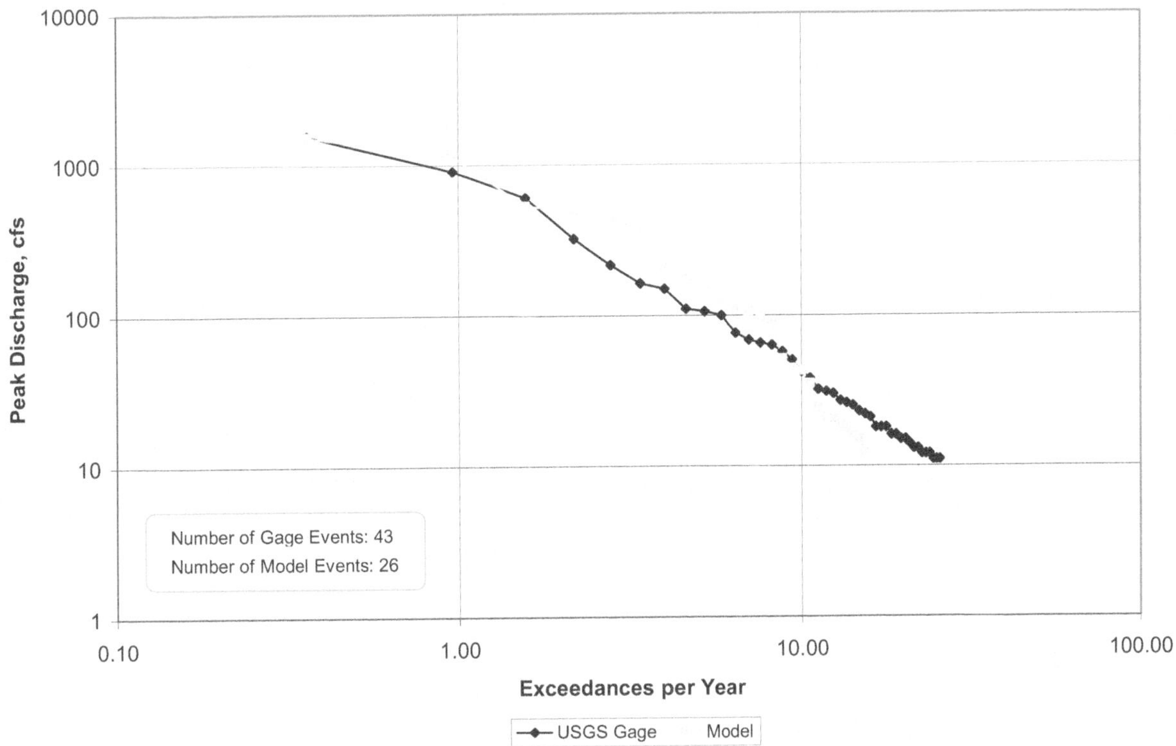

Figure 4-4. Morgan Watershed Peak Flow Frequency Exceedance Calibration.

SWMM models at 15-minute time intervals. This metric, modeled after Booth et al. (2004), is a measure of stream flashiness and in this analysis is evaluated as *the percentage of time flow is above the peak discharge of a half-year return interval storm*. The $T_{0.5}$ hydrologic metric was evaluated for six time periods:

- the 12–17 month period of record surrounding the biologic data collection, which is the period of level/flow monitoring at each USGS EUSE site, and the period of record the models were calibrated to, ranging from July 2002 to November 2003;
- the 1-year period of record prior to the biologic data collection: July 2002 to June 2003;
- the 2-year record prior to the biologic data collection: July 2001 to June 2003;
- the 5-year record prior to the biologic data collection: July 1998 to June 2003;
- the 10-year record prior to the biologic data collection: July 1993 to June 2003; and
- the 20-year record prior to the biologic data collection: January 1986 to December 2005.

The $T_{0.5}$ hydrologic metric was also computed at hourly and daily time steps to determine the sensitivity of the metric to temporal resolution. It was found that the $T_{0.5}$ hydrologic metric computed from an hourly time step correlated with biologic data similarly to the $T_{0.5}$ hydrologic metric values computed from a 15-minute time step. The $T_{0.5}$ hydrologic metric computed from a daily timestep, however, did not correlate well with the biologic data (R^2 range from 20-year flow record: 0.0014–0.076), and the difference in the metric values computed for each watershed from the daily timestep was minimal (for 20-year flow record ranged from 0.561 to 0.578). Results computed from 15-minute timesteps only are presented throughout the remainder of this report.

Geomorphic metrics describing cumulative excess shear stress and erosion potential were calculated using SAS and water depth output from the SWMM models at 15-minute time intervals; the methods described in Section 3.4.2.4 were used for these computations. The geomorphic metrics were evaluated for the 1-, 2-, 5-, 10-, and 20-year periods of record described above; however, they were not computed for the USGS EUSE flow monitoring/ calibration period.

4.3 Relationships Between the $T_{0.5}$ Hydrologic Metric, Measures of Urbanization, and Stream Substrate

Data were initially examined to identify if relationships could be seen between the $T_{0.5}$ hydrologic metric and the urban intensity index values calculated by the USGS for each watershed. Figure 4-5 shows regressions of the $T_{0.5}$ values to the urban intensity index (UII) computed by the USGS (McMahon & Cuffney, 2000; Coles et al., 2004; Tate et al., 2005). These

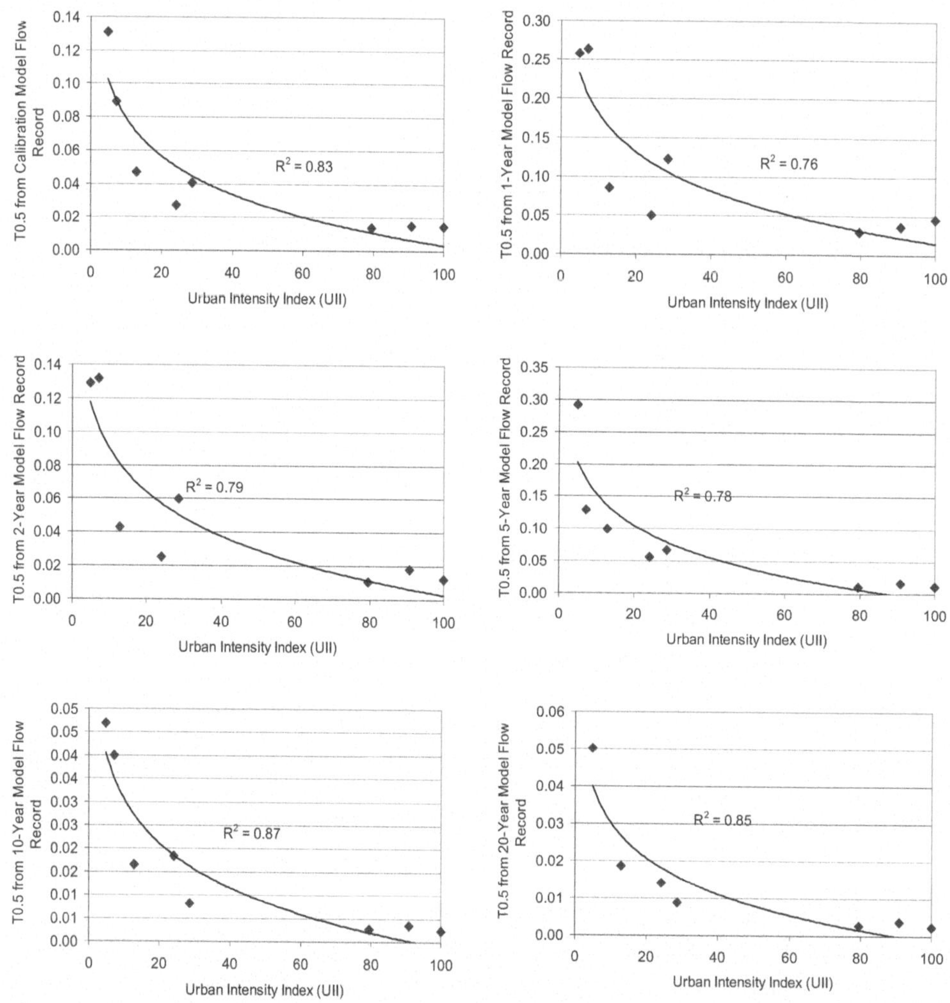

Figure 4-5. Relationship of the Urban Intensity Index to the $T_{0.5}$ Calculated From (a) the Calibration Period, (b) 1-Year Model Flow Record, (c) 2-Year Model Flow Record, (d) 5-Year Model Flow Record, (e) 10-Year Model Flow Record, and the (f) 20-Year Model Flow Record.

figures show a logarithmic relationship between the $T_{0.5}$ and the UII for all six time periods examined, with the strongest relationship occurring during the 10- and 20-year periods.

Because it is known that biological assemblages and individual species are strongly associated with substrate size and condition (Waters, 1995; Berry et al., 2003), the effects of urbanization (UII) and the $T_{0.5}$ hydrologic metric on substrate size were evaluated for the study watersheds. Figure 4-6 shows the relationship between the predominant size class of bed substrate (see Table 4-3) evaluated by the USGS for each sampling location and the UII. The predominant size class was identified for each watershed as the size class that contained the highest percentage of samples with the dominant substrate in that size class. Figure 4-6 shows a logarithmic relationship between predominant size class and the UII.

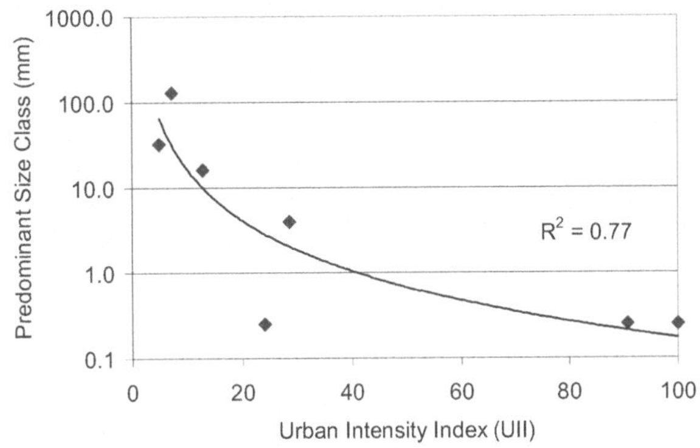

Figure 4-6. Relationship of Predominant Size Class of Substrate to Urban Intensity Index.

Figure 4-7 shows the relationship of the predominant size class of substrate to the $T_{0.5}$ hydrologic metric calculated from (a) the calibration period, (b) 1-year model flow record, (c) 2-year model flow record, (d) 5-year model flow record, (e) 10-year model flow record, and (f) 20-year model flow record. Power function relationships were observed, with the best relationships occurring between the predominant size class and the $T_{0.5}$ hydrologic metric calculated from the calibration and the 1- and 2-year time periods.

4.4 Establishing Linkages Between Hydrologic and Geomorphic Metrics and Biologic Metrics

4.4.1 Relationships Between the $T_{0.5}$ Hydrologic Metric and Macroinvertebrate Metrics

Figures 4-8 show regressions of the $T_{0.5}$ values (computed from modeling results for the two years prior to biologic sampling) to results of the benthic macroinvertebrate sampling conducted by the USGS in the pilot watersheds. Results for EPT richness, a count of the total EPT specimens collected for both RTH and QMH, are displayed in addition to results for EPT percent richness, which is the percent of EPT taxa collected in relationship to the total number of taxa collected. Benthic index of biotic integrity (B-IBI) results are also regressed against $T_{0.5}$ values. The B-IBI values were computed using methods described by Kerans (1994) for rivers of the Tennessee Valley. A B-IBI method is not available specifically for the North Carolina Piedmont ecoregion.

Overall, relationships between the $T_{0.5}$ values and the macroinvertebrate data were strongest for $T_{0.5}$ values computed from the 1-, 2-, and 10-year flow records. R^2 values from the regression for each time period analyzed are summarized in Table 4-10. Regression plots for $T_{0.5}$ values from each time period analyzed are provided in the Appendix.

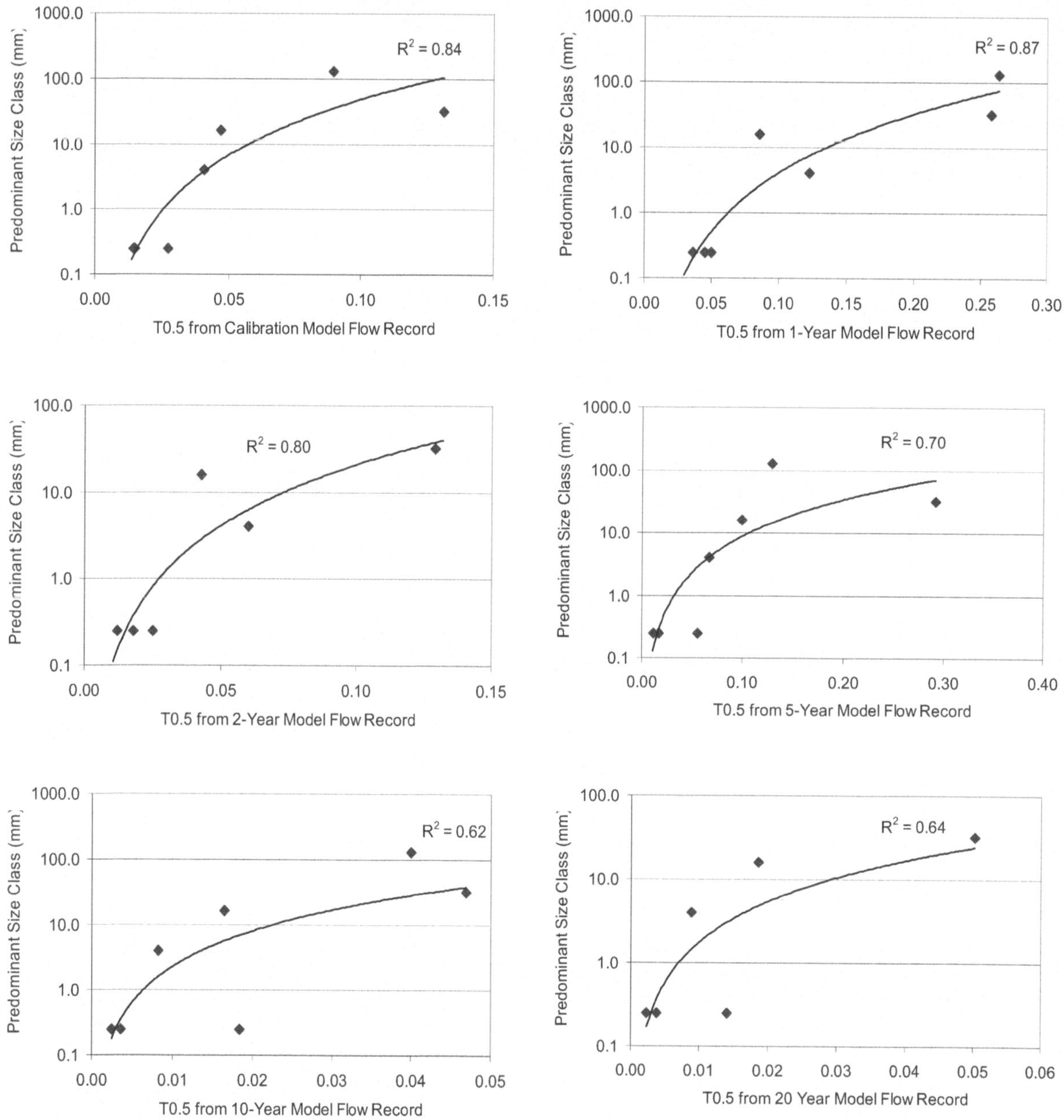

Figure 4-7. Relationship of Predominant Size Class of Substrate to $T_{0.5}$ Calculated From (a) the Calibration Period, (b) 1-Year Model Flow Record, (c) 2-Year Model Flow Record, (d) 5-Year Model Flow Record, (e) 10-Year Model Flow Record, and (f) 20-Year Model Flow Record.

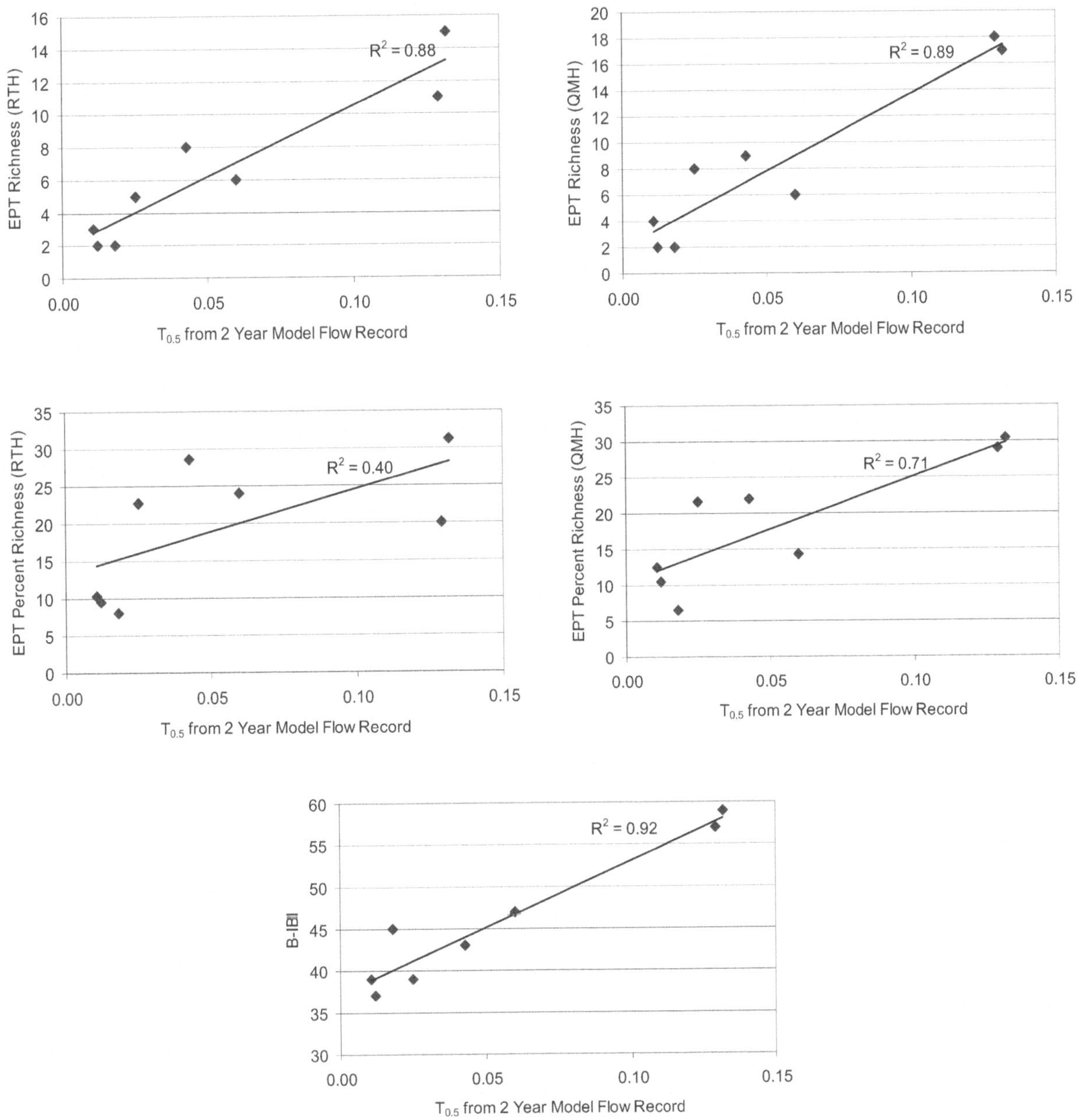

Figure 4-8. Relationship of $T_{0.5}$ Calculated From the 2-Year Flow Record to Benthic Macroinvertebrate Sampling Data (a) EPT Richness (RTH), (b) EPT Richness (QMH), (c) EPT Percent Richness (RTH), (d) EPT Percent Richness (QMH), (e) Benthic Index of Biotic Integrity (B-IBI).

Table 4-10. Summary of Regression Results for Macroinvertebrate Metrics and $T_{0.5}$/Land Use Metrics.

Macroinvertebrate Metric	R^2 Value								
	$T_{0.5}$							Land Use	
	Calib-ration Period	1-Year	2-Year	5-Year	10-Year	20-Year	Mean	UII	DCIA
EPT Richness (RTH)	0.74	0.86	0.88	0.56	0.82	0.69	0.76	0.71	0.57
EPT Richness (QMH)	0.90	0.87	0.89	0.79	0.97	0.91	0.89	0.72	0.53
EPT % Richness (RTH)	0.30	0.38	0.40	0.22	0.38	0.29	0.33	0.84	0.36
EPT % Richness (QMH)	0.72	0.69	0.71	0.64	0.87	0.77	0.73	0.80	0.37
B-IBI	0.78	0.91	0.92	0.61	0.73	0.67	0.77	0.42	0.70
Mean R^2	0.69	0.74	0.76	0.56	0.75	0.67	0.70	0.70	0.51

The weak correlation with $T_{0.5}$ values from the 5-year time period can possibly be explained by the fact that Hurricane Floyd occurred in 1999, which was within the 5-year time period. Hurricane Floyd was a category 4 (very strong) hurricane on the Saffir-Simpson Hurricane Scale (State Climate Office of North Carolina, 2007) that contributed 5.79 inches of rain in a 27-hour time period at the National Weather Service Raleigh Durham International Airport (Coop ID 317069) rain gage. Another possible explanation for weaker correlations during the 5- and 20-year time periods is that average annual precipitation values for these time periods were less than those for the 1-, 2- and 10-year time periods. Variations in precipitation affect the frequency, magnitude, and duration of flows in streams, which, in turn, affect the $T_{0.5}$ hydrologic metrics. Further examination of the effects of precipitation variability on the $T_{0.5}$ hydrologic metric were not explored in this analysis but might be warranted.

In general, the strongest relationships between biologic results and $T_{0.5}$ values were observed for EPT richness (RTH) and for B-IBI. RTH had the weakest correlations in all time periods examined. This weak correlation is attributed to the macroinvertebrate score for the Morgan Creek watershed, which appears to be an outlier with respect to corresponding $T_{0.5}$ values. Interestingly, this value does not appear to be an outlier when plotted against the UII (Figure 4-9).

Figure 4-9 shows a negative correlation between the macroinvertebrate metrics and the UII, with the strongest relationships occurring between the EPT percent richness metrics and a low correlation between B-IBI and UII. Relationships between the macroinvertebrate metrics and DCIA were also examined since imperviousness has historically been indicated as a predictor of aquatic community health (Center for Watershed Protection, 2002). Figure 4-10 shows that the EPT richness metrics weakly correlated with DCIA; however, there was a stronger relationship between DCIA and B-IBI.

It is important to note that although relationships between macroinvertebrate metrics and measures of urbanization such as the UII and DCIA exist, these relationships offer no guidance on opportunities to prevent degradation of biotic integrity of streams short of limiting the extent of urbanization in a watershed. Evaluation of the relationships between macroinvertebrate metrics and hydrologic metrics such as the $T_{0.5}$ allows investigators to identify flow regimes that are conducive to the desired biotic outcome and offers the opportunity to identify stormwater management practices that will produce the desired flow regime. The effects of stormwater

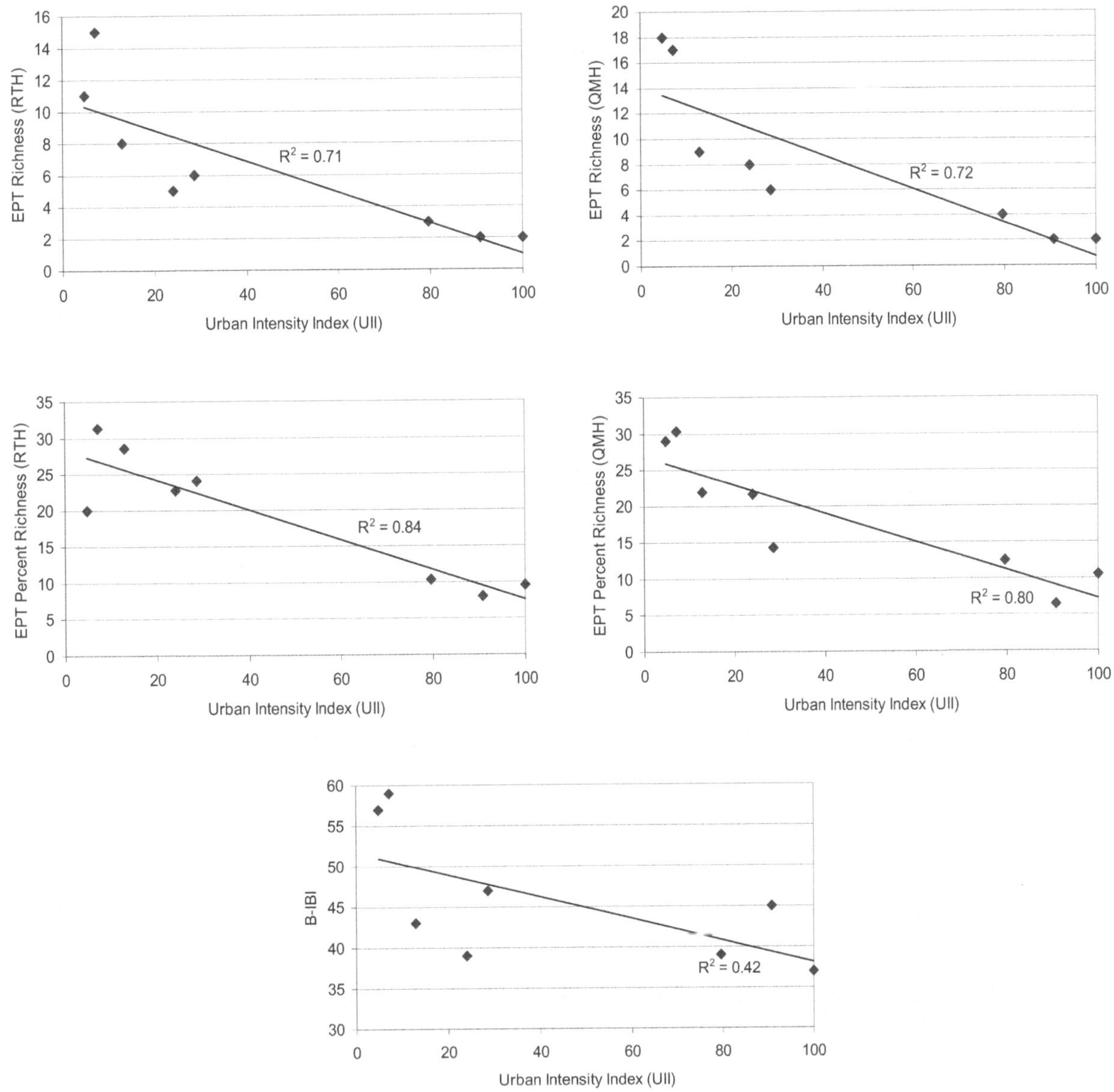

Figure 4-9. Relationship of the Urban Intensity Index to Benthic Macroinvertebrate Sampling Data (a) EPT Richness (RTH), (b) EPT Richness (QMH), (c) EPT Percent Richness (RTH), (d) EPT Percent Richness (QMH), (e) Benthic Index of Biotic Integrity (B-IBI).

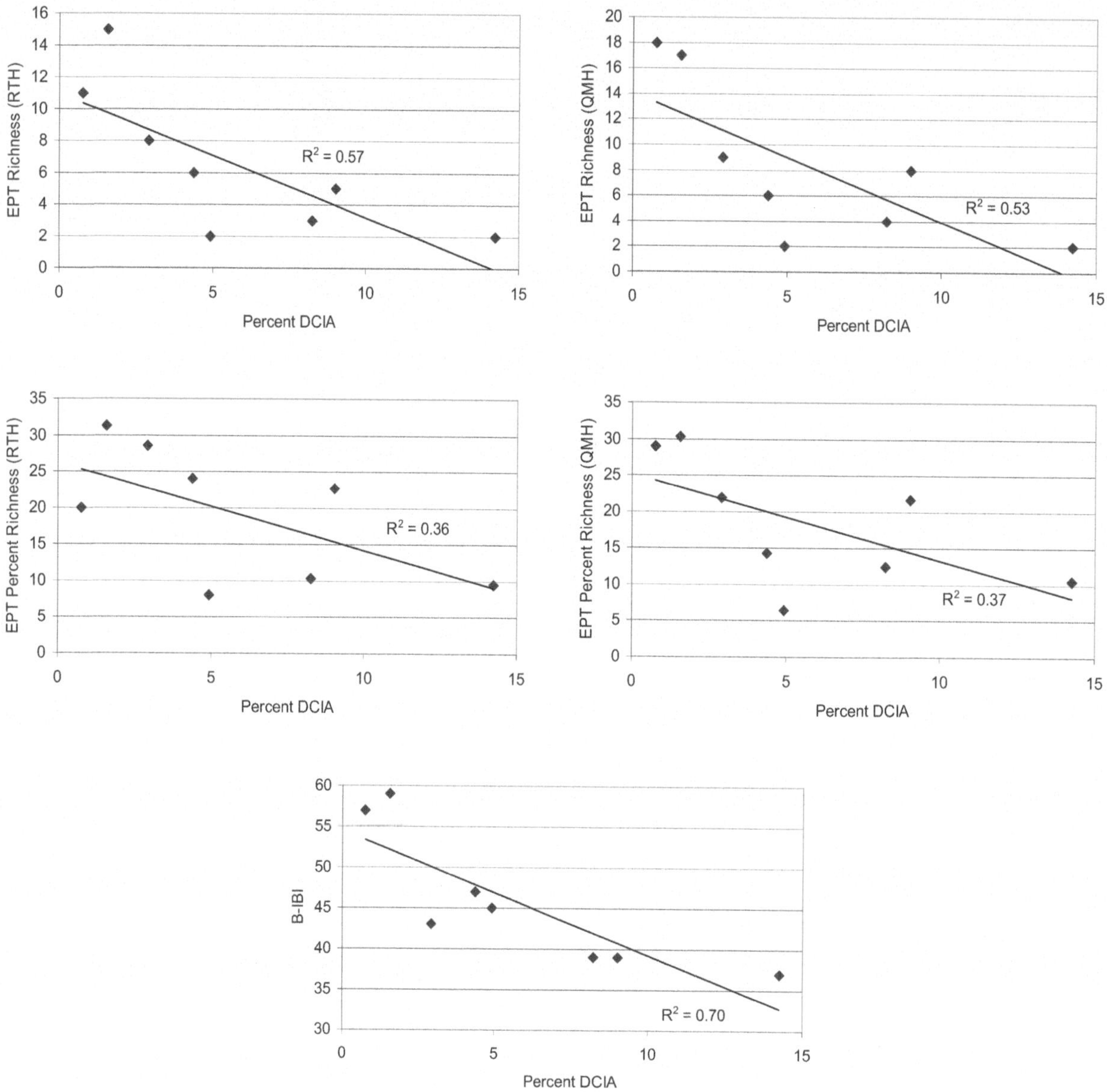

Figure 4-10. Relationship of Percent DCIA to Benthic Macroinvertebrate Sampling Data (a) EPT Richness (RTH), (b) EPT Richness (QMH), (c) EPT Percent Richness (RTH), (d) EPT Percent Richness (QMH), (e) Benthic Index of Biotic Integrity (B-IBI).

management practices on the $T_{0.5}$ metric in the North Carolina Piedmont is discussed in Section 4.5.1.

Results from the EPT richness (QMH) and the $T_{0.5}$ metrics regression analyses for the 10-year time period combined with the NCDENR (2006) EPT richness rating scores shown in Table 4-2 were used to identify $T_{0.5}$ values that would create flow conditions favorable to desirable EPT richness rating scores. An example of this process is demonstrated in Figure 4-11.

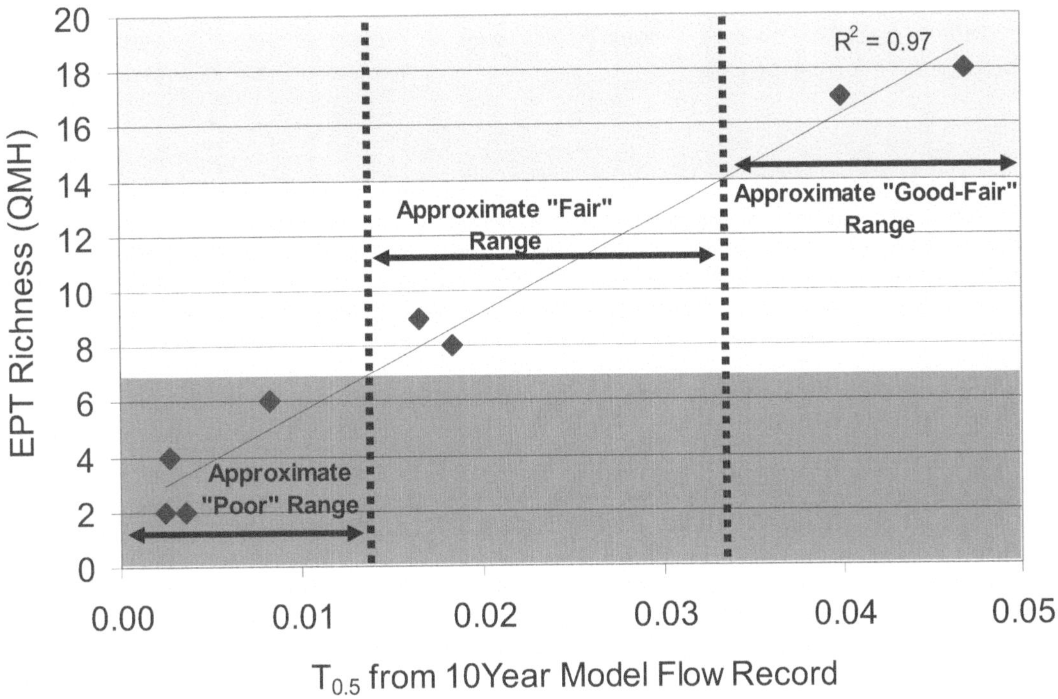

Figure 4-11. Relating $T_{0.5}$ Values for the 10-Year Flow Record to EPT Richness Rating Scores.

4.4.2 Relationships Between Geomorphic Metrics and Biologic Parameters

Figure 4-12 relates the predominant size class identified for each watershed to the corresponding macroinvertebrate score. Predominant categories range from sand to large cobble. As expected, the biologic scores are positively, logarithmically related to substrate size, with EPT richness for the RTH having the strongest correlation with an R^2 of 0.90.

Figure 4-13 shows the relationship of the macroinvertebrate metrics to the cumulative excess shear stress computed for the 1-year period of record prior to the benthic macroinvertebrate sampling. The cumulative excess shear stress over the 1-year simulation period was calculated for the grain size of the predominant bed substrate at the sampling site of each watershed. Cumulative excess shear stress was not calculated for the Black Creek watershed because the predominant class that includes large boulders, irregular bedrock, irregular hardpan, or irregular artificial surface does not have a definitive grain size. During the modeled 20-year time period, no flows occurred in the Black Creek watershed that would move sediment with a grain size greater than 512 mm. It should also be noted that no flows occurred in the Strouds Creek watershed that were large enough to entrain the predominant substrate in that watershed, which was the size class large cobble with a d_s > 128 mm.

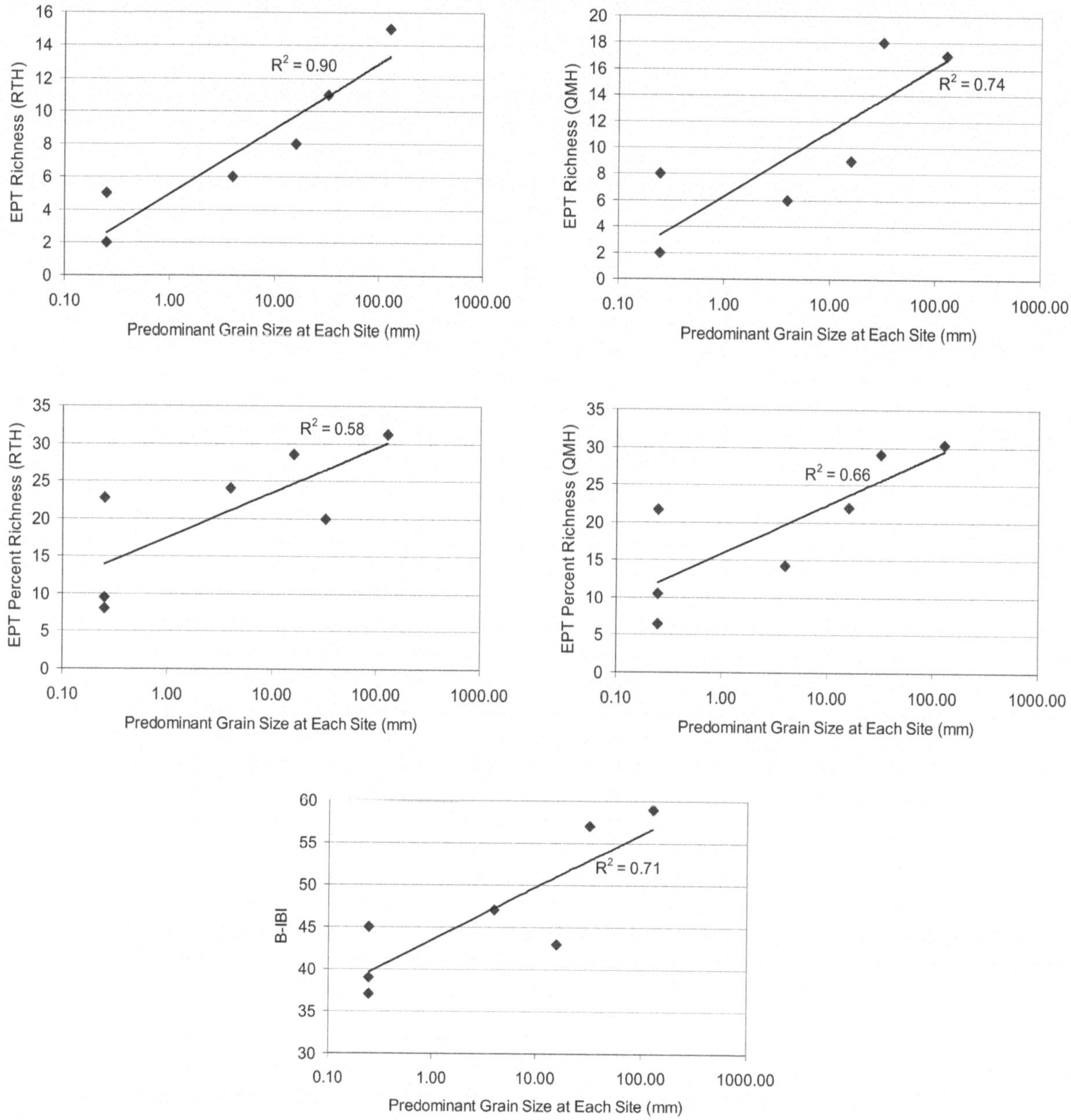

Figure 4-12. Relationship of Predominant Grain Size at Each Site to (a) EPT Richness (RTH), (b) EPT Richness (QMH), (c) EPT Percent Richness (RTH), (d) EPT Percent Richness (QMH), (e) Benthic Index of Biotic Integrity (B-IBI).

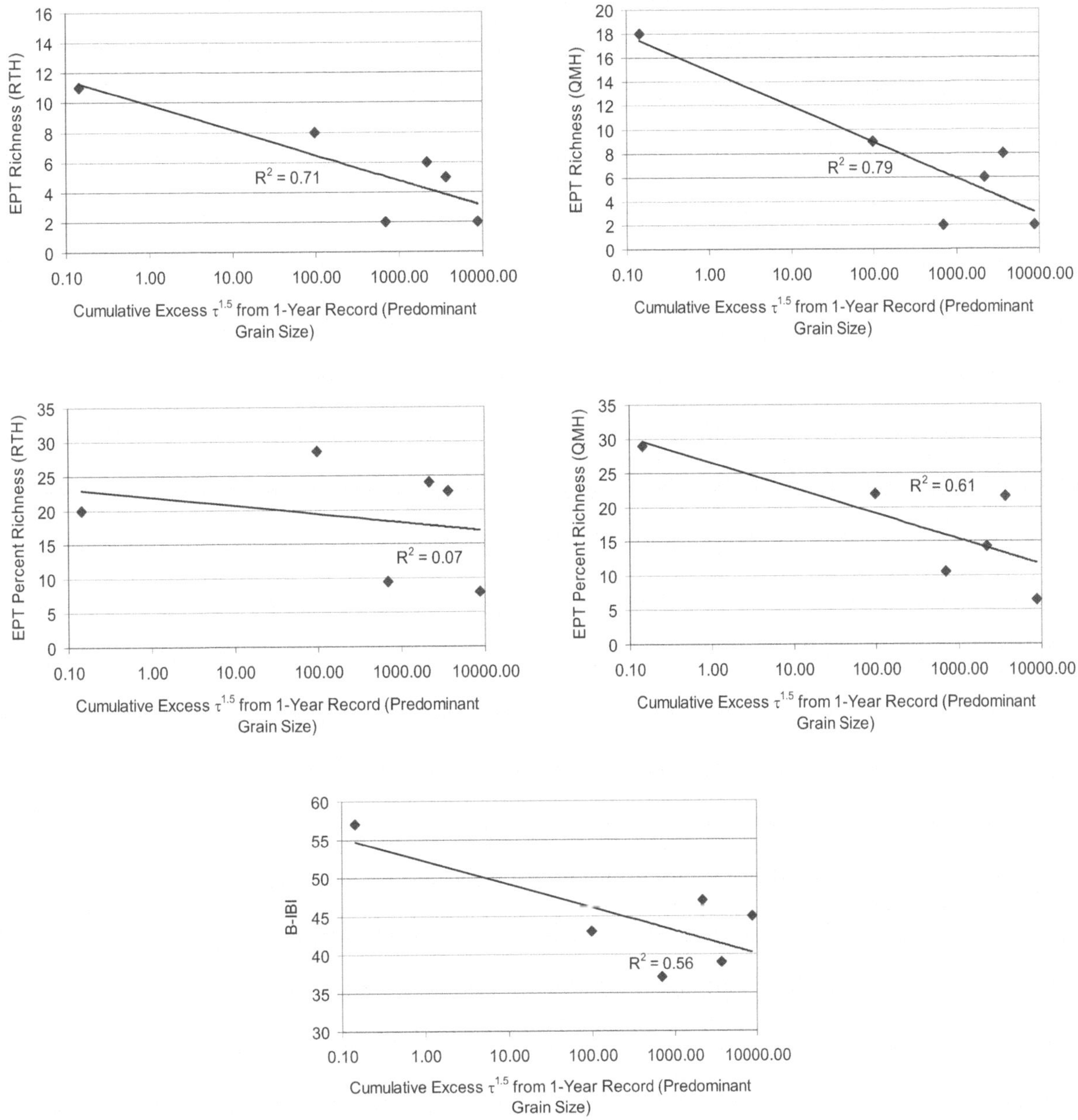

Figure 4-13. Relationship of Cumulative Excess Shear from the 1-Year Flow Record to Benthic Macroinvertebrate Sampling Data (a) EPT Richness (RTH), (b) EPT Richness (QMH), (c) EPT Percent Richness (RTH), (d) EPT Percent Richness (QMH), (e) Benthic Index of Biotic Integrity (B-IBI).

Similar to the results relating predominant substrate to macroinvertebrate metrics, EPT richness showed the strongest relationships between cumulative excess shear stress from the 1-year time period examined and the benthic macroinvertebrate metrics. This is likely due to the influence of the substrate on habitat quality.

When compared to the macroinvertebrate metrics, cumulative excess shear stress values computed for the 2-, 5-, 10- and 20-year time periods yielded similar regression results to those for the one-year time period. A summary of the R^2 values for each time period is provided in Table 4-11. Table 4-11 also summarizes the regression results for the macroinvertebrate metrics and the predominant substrate for each watershed shown in Figure 4-12. Overall, the predominant substrate for each watershed was a better predictor of macroinvertebrate response than was cumulative excess shear stress.

Table 4-11. Summary of Regression Results Between Macroinvertebrate Metrics and Geomorphic Metrics.

Macroinvertebrate Metric	R^2 Value						
	Cumulative Excess $\tau_{1.5}$						Predominant Substrate
	1-year	2-year	5-year	10-year	20-year	Average	
EPT Richness (RTH)	0.69	0.69	0.69	0.69	0.69	0.69	0.90
EPT Richness (QMH)	0.78	0.77	0.76	0.75	0.75	0.76	0.74
EPT % Richness (RTH)	0.06	0.06	0.08	0.07	0.07	0.07	0.58
EPT % Richness (QMH)	0.59	0.58	0.60	0.59	0.58	0.59	0.66
B-IBI	0.56	0.55	0.50	0.50	0.53	0.53	0.71
Mean R^2	0.54	0.53	0.53	0.52	0.52	0.53	0.72

Because the critical shear stress value assigned to each individual stream was determined by the grain size of the bed substrate, cumulative excess shear stress is not independent of the size of the bed substrate. From this it follows that the response of the macroinvertebrates to cumulative excess shear cannot be considered independently of the bed substrate. Because substrate was independently a better predictor of macroinvertebrate response than cumulative excess shear, this research shows that predominant substrate should be used as a geomorphic metric to predict macroinvertebrate response rather than cumulative excess shear. The computation of cumulative excess shear, however, is necessary for determining erosion potential, which will be shown to be a useful tool in evaluating stormwater management scenarios in Section 4.5.2.

The relationship between cumulative excess shear stress and the predominant size class used to compute the cumulative excess shear stress is shown in Figure 4-14. Relationships between cumulative excess shear stress and the predominant size class were nearly the same for each time period examined and were negatively related on a log-log scale.

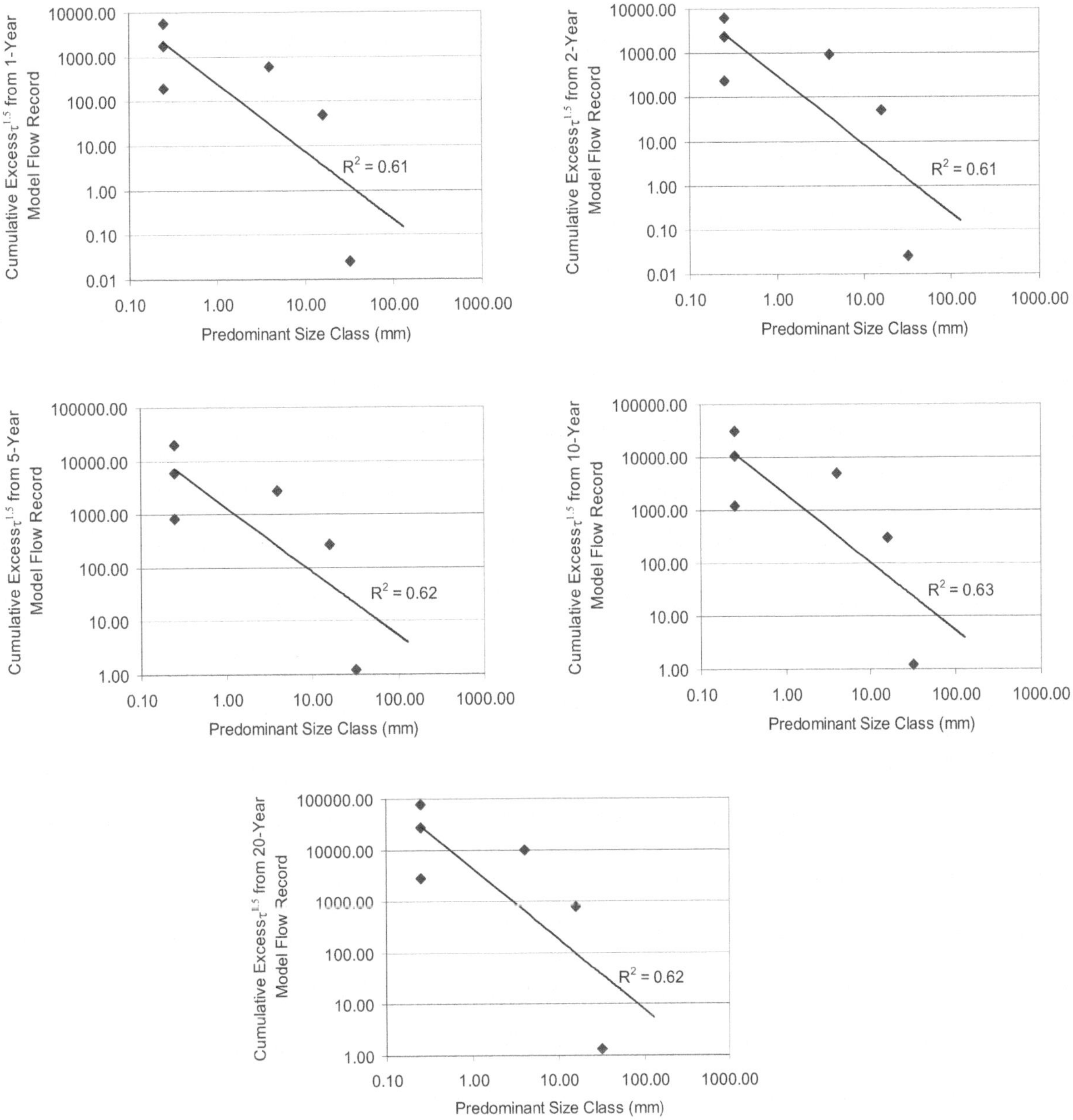

Figure 4-14. Relationship of the Predominant Size Class to Cumulative Excess Shear from the (a) 1-Year Model Flow Record, (b) 2-Year Model Flow Record, (c) 5-Year Model Flow Record, (d) 10-Year Model Flow Record, and (e) 20-Year Model Flow Record.

4.5 Evaluating the Impact of Land Use Change and Runoff Control Strategies on Hydrologic and Geomorphic Metrics

The effects of land use and channel geometry on hydrologic and geomorphic metrics were evaluated in the Morgan Creek watershed. Four scenarios were evaluated:

1. Urbanized land use in the watershed with urbanized channel geometry,
2. Urbanized land use in the watershed with rural channel geometry,
3. Rural land use in the watershed with urbanized channel geometry, and
4. Rural land use in the watershed with rural channel geometry (existing conditions).

Urbanized land use was based upon existing land use in the Pigeon House Creek watershed, which has an overall average of 14.3% DCIA. A uniform DCIA value of 14.3% was applied to all subcatchments in the Morgan watershed for the urbanized land use scenarios. Rural land use was modeled using existing land use in the watershed, calculated as described in Section 4.2.1, and varied by subcatchment from 0% to 11.8% DCIA, with an overall average of 0.8% DCIA in the watershed. Channel geometry for rural (existing) conditions in the Morgan Creek watershed was based upon field measurements described in Section 4.2.1. Channel geometry for urbanized conditions was based on typical geometry corresponding to respective drainage areas in the Pigeon House Creek watershed. For example, the urbanized channel geometry applied in the Morgan Creek watershed was derived from the channel geometry representing a similar drainage area in the Pigeon House Creek watershed. Examples of representative cross-sections for urbanized and rural channel geometry for the Morgan Creek watershed are shown in Figure 4-15.

Stormwater best management practices (BMPs) for flood control and water quality treatment were applied to the urbanized scenarios in the Morgan Creek watershed (14.8% DCIA). Twenty-seven detention basins were modeled (Figure 4-16) to accomplish control of the water quality capture volume (WQCV) and control of the 2-year and 10-year design storms to the peak discharge values observed under rural watershed conditions (0.8% DCIA). The WQCV was computed as the product of the mean runoff event and a drawdown coefficient, resulting in the capture of 70% to 90% of all runoff-producing events in their entirety. Using procedures recommended by the Water Environment Federation and the American Society of Civil Engineers (1998) and a draw-down time of 40 hours, the resulting WQCV was calculated as:

$$P_o = (ac)P_6 \qquad (4.4\text{-}1)$$

where a = drawdown coefficient (1.299), P_6 = mean storm precipitation over the watershed (18.0 mm [WEF & ASCE, 1998]) and c = runoff coefficient. The runoff coefficient was calculated using:

$$c = 0.858i^3 - 0.78i^2 + 0.774i + 0.04 \qquad (4.4\text{-}2)$$

where i = watershed imperviousness ratio. Orifice outlets for the 2-year and 10-year peak shaving were sized using an iterative procedure in Microsoft Excel spreadsheet software. All detention basins were 11-ft deep or less and had orifices varying in size from 0.5 ft to 2.5 ft in diameter. Basins were modeled with square footprints for simplicity. Two general outlet designs are illustrated in Figure 4-17.

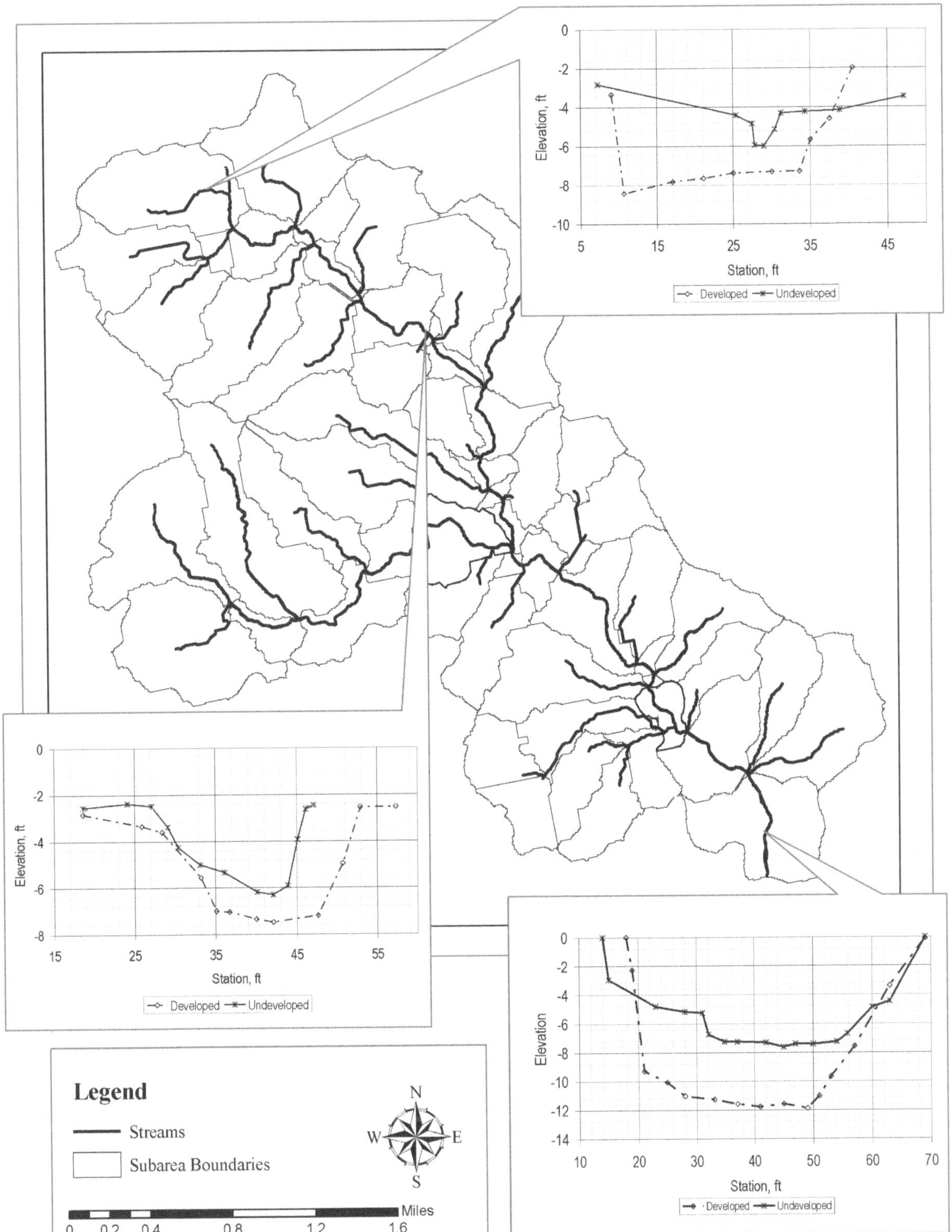

Figure 4-15. Representative Cross-Sections for Rural and Urbanized Channels in the Morgan Creek Watershed.

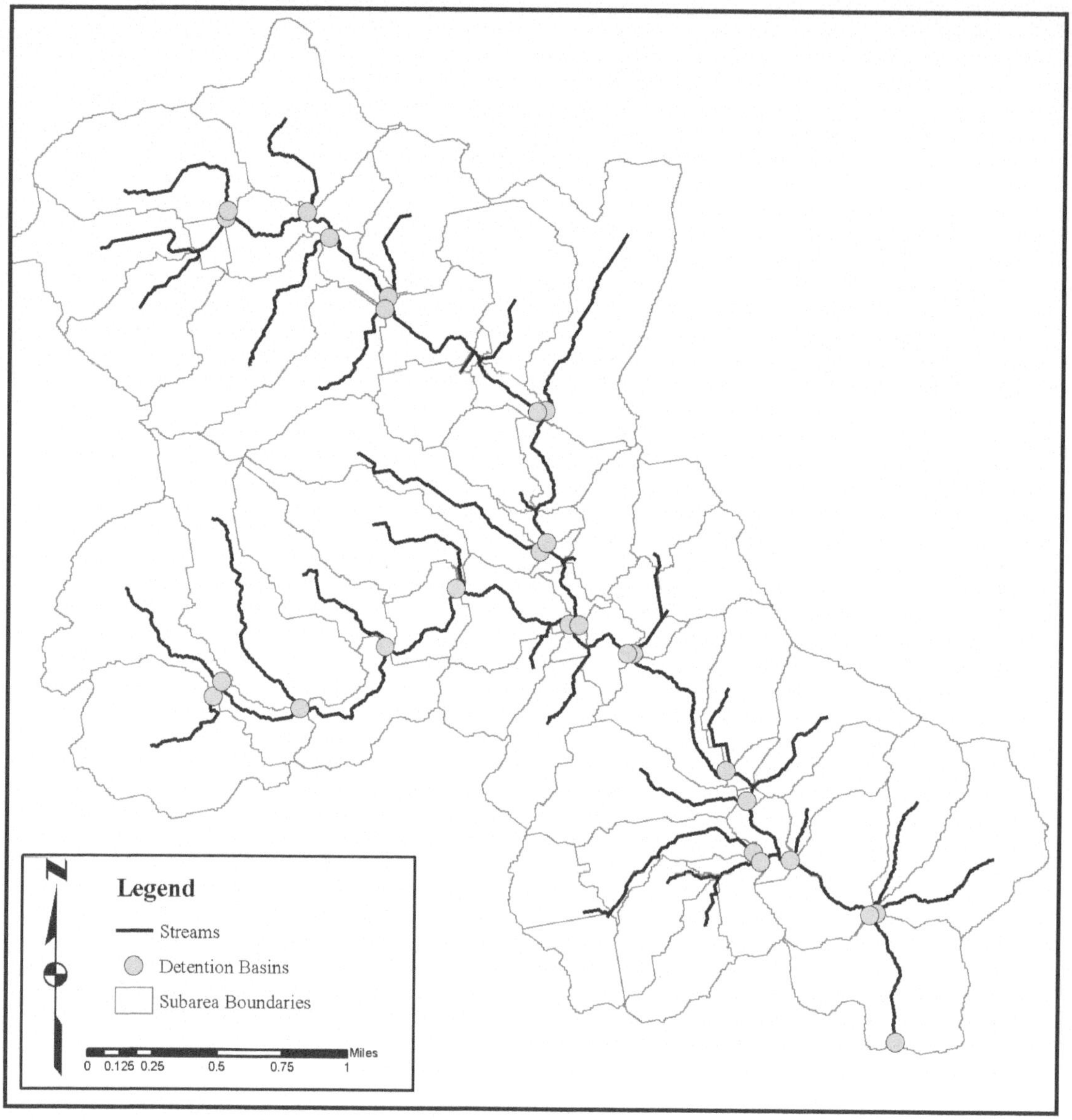

Figure 4-16. Location of Detention Basins Modeled in the Morgan Creek Watershed.

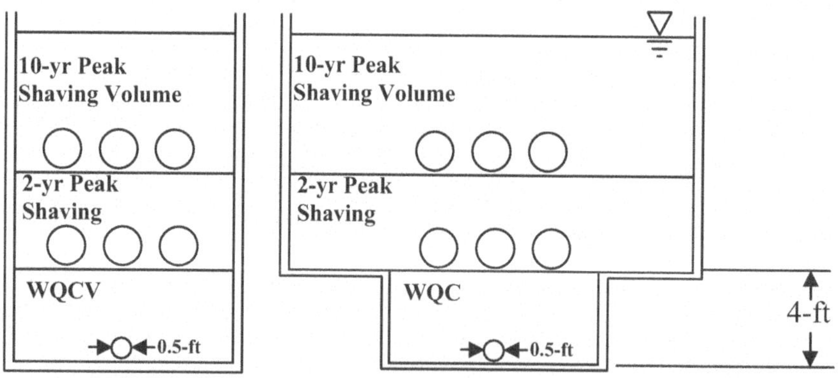

Figure 4-17. General Design of Stormwater Detention Pond Outlets.

WERF

4.5.1 Hydrologic Analysis

As described in the previous section, the effects of land use and channel geometry on hydrologic and geomorphic metrics were evaluated in the Morgan Creek watershed. This section illustrates the hydrologic effects of land use changes, channel geometry changes, and the application of stormwater best management practices using peak flow frequency exceedance curves, flow duration curves, and the $T_{0.5}$ hydrologic metric.

Figures 4-18 and 4-19 display the peak flow frequency exceedance curves for all of the scenarios evaluated in the Morgan Creek watershed. Existing conditions in the watershed are illustrated by the line depicting rural land use and rural channel geometry in Figure 4-18. As anticipated, the lowest peak discharge values occur under the rural land use, rural channel condition. The change in land use in the watershed from urban to rural creates an increase in peak discharge values across the full range of return periods when both rural and urban channels are simulated; a significantly larger increase in peak discharge values is observed when urban land use is combined with the urban channels. These increases in peak discharge are due to increased conveyance capabilities throughout the system. When channels are widened and deepened, upstream attenuation of peak flows is reduced with decreases of overbank flooding and the loss of storage provided through floodplain connectivity. Figures 4-18 and 4-19 also show the cumulative downstream effects of the 27 stormwater BMPs applied to the Morgan Creek watershed. Figure 4-19 demonstrates that the stormwater BMPs were designed to control peak discharges for rural channel geometry because the peak discharges are far below those for rural land use and urban channel geometry conditions.

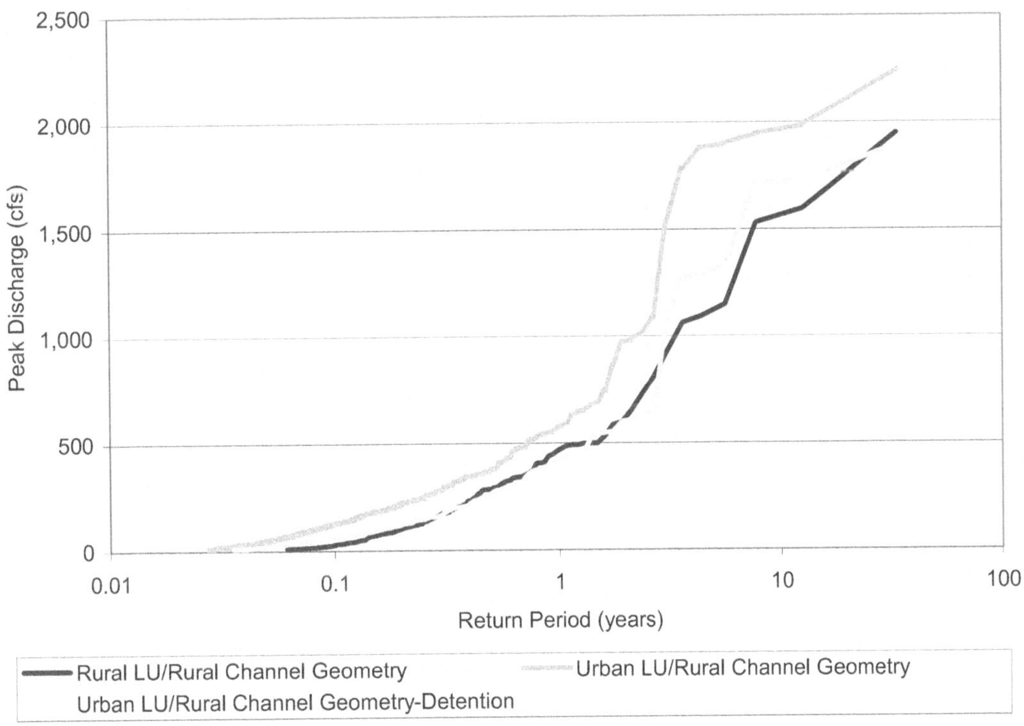

Figure 4-18. Effects of Detention on Peak Flow Frequency Exceedance Curves in Morgan Watershed with Rural Channels.

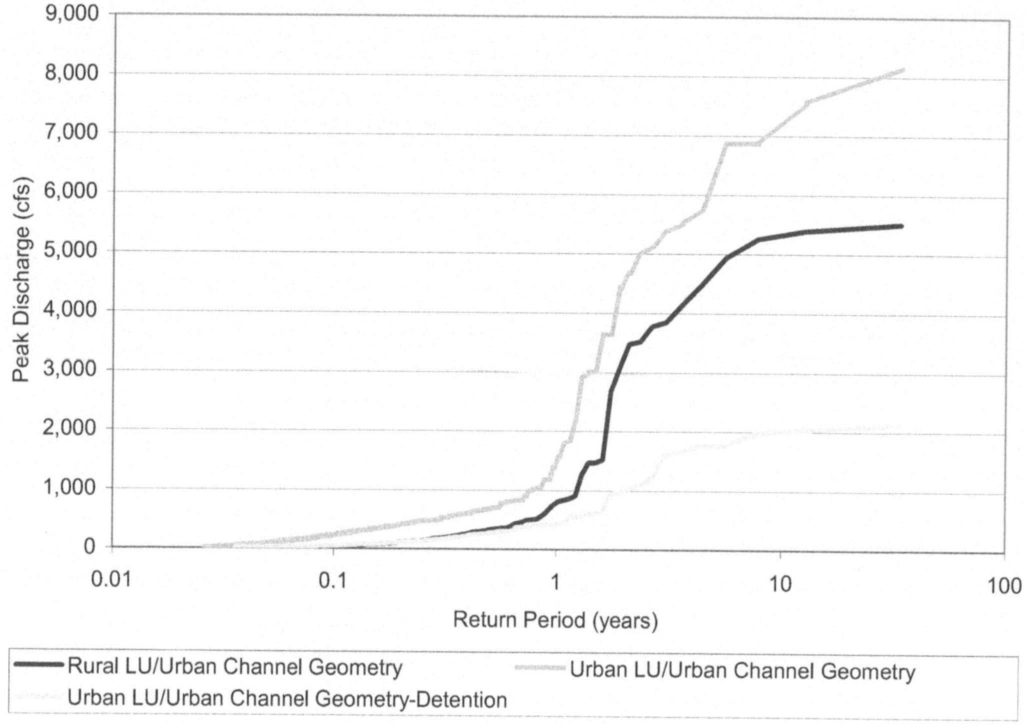

Figure 4-19. Effects of Detention on Peak Flow Frequency Exceedance Curves in Morgan Watershed with Urban Channels.

Figures 4-20 and 4-21 display the effects of land use change, channel geometry change, and stormwater management practices on the full spectrum of flows, as demonstrated by flow duration curves. Figure 4-20 shows that when runoff generated under urban conditions is controlled by stormwater BMPs, the magnitude and duration of large discharges (those that occur less than 1% of the time) closely matches those of rural conditions across a full spectrum of discharges. Changes in the durations of low flows due to the extended release of stormwater from the BMPs are not easily observed at the outlet of the watershed due to contributions of subsurface flows to baseflow in the model.

The $T_{0.5}$ hydrologic metric was computed for the four combinations of land use and channel geometry conditions modeled in the Morgan Creek watershed, as shown in Figure 4-22. Results were computed for the calibration period, 1-year model flow record, 2-year model flow record, 5-year model flow record, 10-year model flow record, and 20-year model flow record. The effects of stormwater management practices on the $T_{0.5}$ are also shown.

Ranges of $T_{0.5}$ metric values for each of the three categories of EPT richness scores (poor, fair, and good-fair) were identified from the regression plots for each time period examined. Figure 4-11 illustrates this process. The colored panels in Figure 4-22 indicate the target ranges. It is important to note that the colored panels represent target ranges and are not meant to specifically identify thresholds, as there is a great deal of uncertainty associated in the modeling efforts that were used to compute the $T_{0.5}$ metric values. Additionally, a number of factors other than the hydrologic flow regime play into macroinvertebrate health, such as water quality; these factors are not included in this analysis. Meeting the target ranges of $T_{0.5}$ metric values may not allow the target macroinvertebrate scores to be reached if other factors are at play.

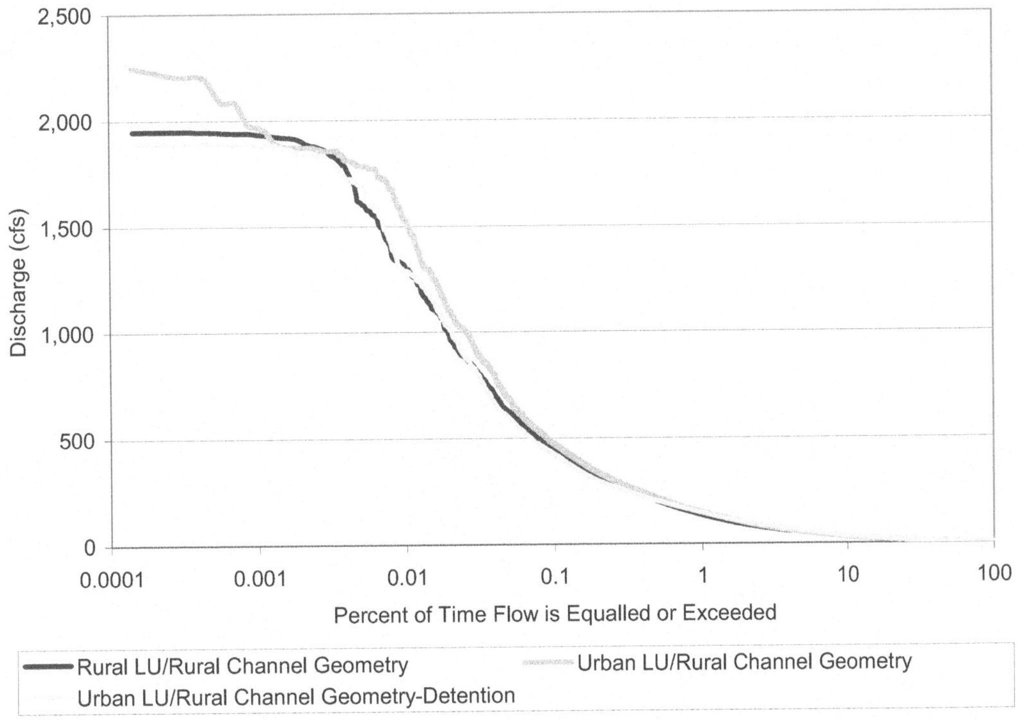

Figure 4-20. Effects of Detention on Flow Duration Curves in Morgan Watershed with Rural Channels.

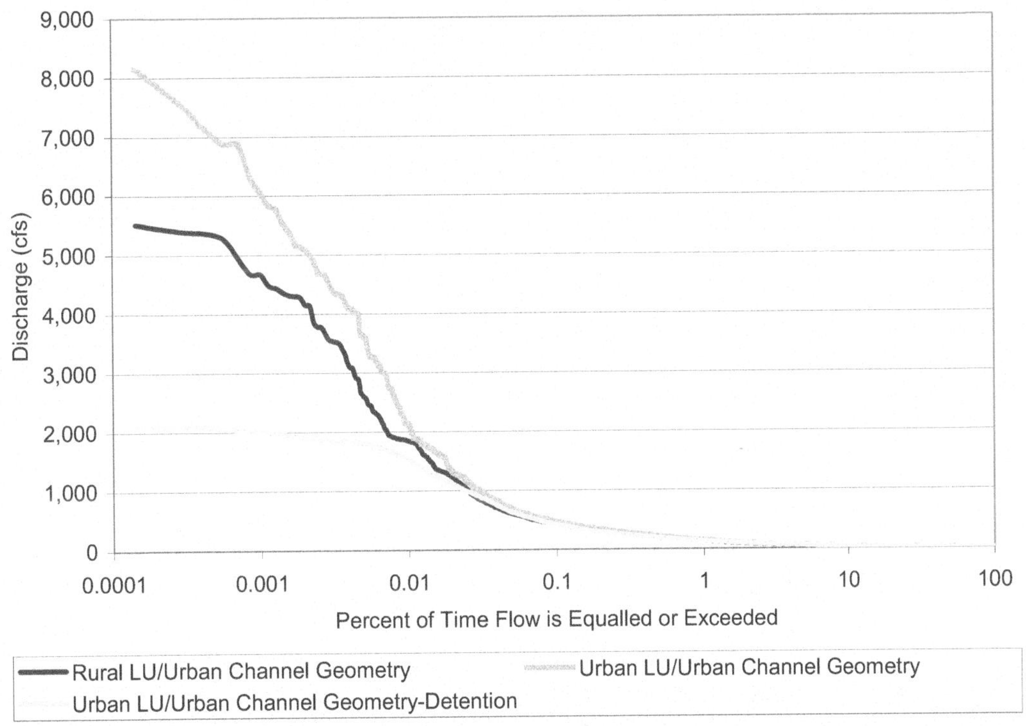

Figure 4-21. Effects of Detention on Flow Duration Curves in Morgan Watershed with Urban Channels.

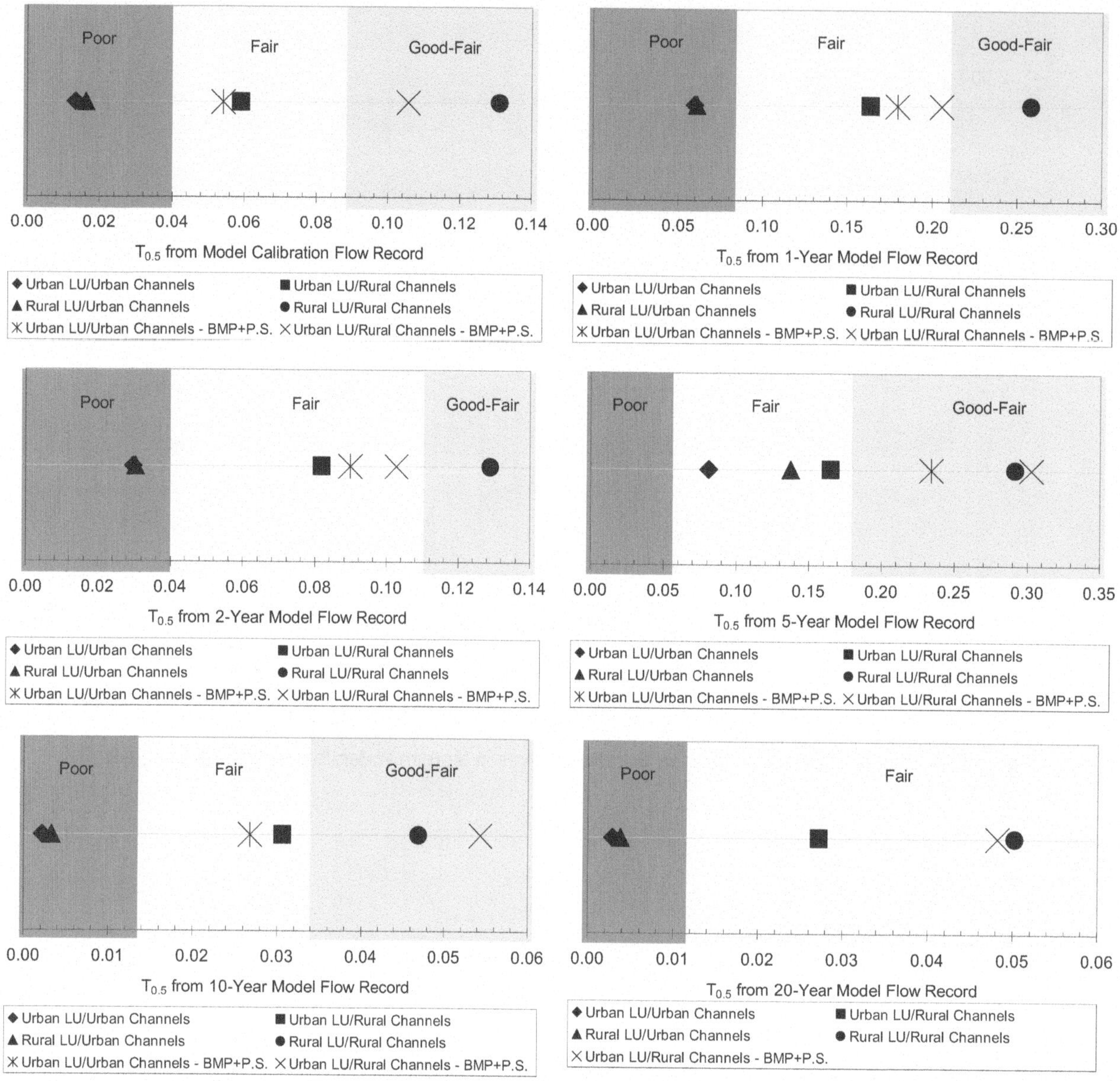

Figure 4-22. Effects of Land Use Change, Channel Geometry Change, and Stormwater Management Practices on the $T_{0.5}$ Metric in the Morgan Creek Watershed From the (a) the Calibration Period, (b) 1-Year Model Flow Record, (c) 2-Year Model Flow Record, d) 5-Year Model Flow Record, (e) 10-Year Model Flow Record, and (f) 20-Year Model Flow Record.

Figure 4-22 shows that channel geometry plays a significant role in the flashiness of stream flows, as demonstrated by the $T_{0.5}$ metric values. In both urban and rural land use conditions, the loss of upstream flow attenuation creates flow conditions that could cause degradation in EPT richness. Consistent with the findings of Edgerly (2006), Figure 4-22 shows that application of stormwater BMPs in the Morgan watershed affects the $T_{0.5}$ metric. The stormwater BMPs prevent the $T_{0.5}$ metric from decreasing to a value that would be in the range indicative of a poor EPT richness rating. The highest $T_{0.5}$ values were obtained when stormwater BMPs were applied in the watershed modeled with rural channels. This demonstrates the importance of preventing stream channel erosion and maintaining existing channel geometries when urbanization of a watershed takes place.

4.5.2 Geomorphic Analysis

This section demonstrates the influence of land use and stormwater management practices on instream erosion potential in the study watersheds. A full description of the methods used to compute erosion potential is given in Section 3.4.2.4.

Table 4-12 shows the effect of stormwater management practices on erosion potential using cumulative excess shear stress values to the 1.5 power for very coarse gravel ($d_s > 32$ mm). The predominant size classes in the Morgan Creek watershed are very coarse gravel and small cobble ($d_s > 64$ mm), each class individually accounted for 22% of the samples evaluated. The erosion potential analysis was conducted for rural channels, only, as it is not appropriate to relate cumulative excess shear stress values computed for channels with different geometry. Table 4-12 shows that a desired erosion potential value of 1.0 is achieved when erosion potential is computed for the 20-year period of record; a slight over-control of erosion potential occurs for the 5- and 10-year periods of record. This over-control was expected, given that the flow duration curves in Figure 4-20 showed that the high-magnitude flows at the outlet of the Morgan Creek watershed were slightly over-controlled due to the downstream combined effects of the 27 stormwater basins. The high erosion potential values computed for the 1- and 2-year time periods occur because the cumulative excess shear values for those time periods were extremely small because very few flows occurred that were large enough to entrain the substrate. It is more appropriate to evaluate erosion potential for a longer period of time, when a larger spectrum of flows is more likely to occur.

It should be noted that the erosion potential computations do not account for changes in upland sediment supply that occur with land use change in a watershed or the application of stormwater management BMPs. Stream channel stability is dependant on a balance between sediment transport capacity of a stream and the supply of sediment both from upland and in-stream sources, therefore erosion potential values of less than 1.0 do not necessarily mean that the flows in the channel are over-controlled.

Table 4-12. Summary of Erosion Potential for the Morgan Creek Watershed.

Scenario	1-Year	2-Year	5-Year	10-Year	20-Year
Developed, No Controls	12.54	12.54	1.09	1.24	1.54
Developed, BMP+Peak Shaving	4.37	4.37	0.93	0.93	0.99

CHAPTER 5.0

BIOLOGICAL AND HABITAT SAMPLING PROTOCOLS

5.1 Biological Sampling

Pursuant to Steps 2 and 4b of the protocol (see Figure 1-2), biological parameters and sampling locations should be selected for and collected from a given study area based on the management goals that the study is intended to support. Both selecting goals for urban waters and deciding which streams or sample locations to exclude or censor from further study requires a good conceptual model for the range of biological conditions that can be expected in a region. Optimally, study sites should be selected that not only exhibit the full range of development / land use in the study area but that are also representative of a biological condition gradient (BCG) for aquatic life (U.S. EPA, 2005; Barbour & Yoder, 2007) in that region.

The following sections discuss the recommended process for determining a minimum set of biological parameters to include in a study of the biological condition of streams in developed and undeveloped watersheds. Identification of study parameters should be predicated on established data quality objectives (DQOs) that ultimately are tied to the intended use of the study and associated results.

5.1.1 Bioassessment Study Design

Qualitative and quantitative DQOs should be developed in light of the identified ecological goals (Step 1, Figure 1-2) that will drive the choice of indicators/biological parameters to be measured (Step 2, Figure 1-2). According to Barbour and Yoder (2007), the DQO process identifies the goals and the potential consequences for making determinations in error with regard to attainment or nonattainment of specified goals. The DQO process also establishes the level of confidence in making the correct determination (e.g., 90% level of confidence that biological condition at a given site is impaired). To this end, related to the complexity of management situations and key technical elements needed to address this complexity, Barbour and Yoder (2007) have differentiated bioassessment programs into four basic data condition levels, with Level 1 being the most basic and Level 4 being the most rigorous and comprehensive. Increasing the comprehensiveness of a bioassessment generally results in data with greater ability to detect changes along the biological condition gradient (Figure 5-1). The various data condition levels are described below in descending order of comprehensiveness:

- **Level 4** employs numeric biological criteria, which are based on assessment tools such as calibrated indices; e.g., IBI, Modified Index of Well-Being (MIwb) for fish, ICI for aquatic macroinvertebrates (Ohio EPA, 1987; 1989) that are based on well-defined regional reference conditions. While acknowledging that few areas of the United States are completely untouched by human impacts, reference conditions should ideally be established for the least-impacted streams within the geographic region in which the study area is located.
- **Level 3** represents bioassessments employing either numeric biological criteria, narrative criteria, or both. Narrative biological criteria generally categorize observed biological condition using an ordinal scale of narrative descriptors ranging from "poor" to "excellent."

Level 3 assessments are able to discriminate among fewer condition categories than assessments performed at the Level 4 detail.

- **Level 2** bioassessments are of sufficient detail to discriminate only broad differences in biological condition among study sites, essentially supporting "pass/fail" determinations (e.g., benchmarks for volunteer monitoring of waterbodies such as scenic rivers commonly performed by watershed groups).
- **Level 1** bioassessments are limited to assessment of obvious and severe degradation and are best utilized as a general screening tool; as a screening tool they may yield information in support of refining information needs and additional study design at higher levels of data resolution (e.g., bioassessment efforts used by untrained volunteers in monitoring streams or rivers in an educational context).

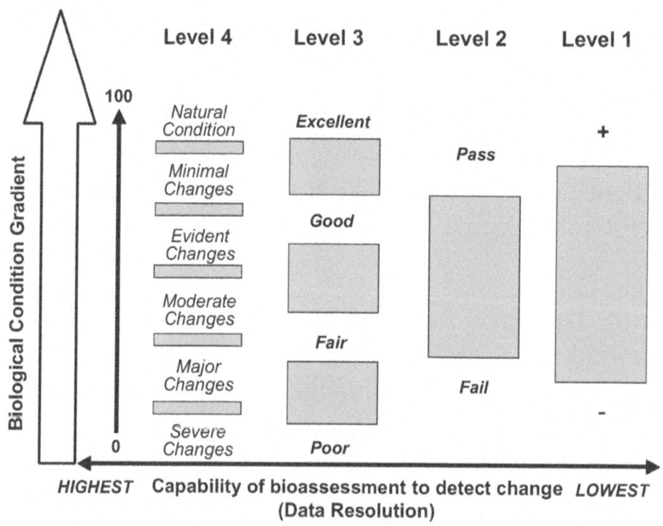

Figure 5-1. Conceptual Illustration of the Capability of Increasingly Comprehensive Bioassessment to Detect Change Along a Biological Condition Gradient (adapted from Barbour & Yoder, 2007).

Increasing the level of study resolution increases the adequacy of data not only for detecting differences in biological condition but also for determining causal associations or stressor-response relationships (Table 5-1). Concerning biological condition assessment, bioassessments at all data condition levels may support some level of determination of impairment versus non-impairment through comparison of observed biological condition to reference conditions or other criteria. Data at higher condition levels may extend beyond determinations of impairment and characterize observed biological condition along a scale of multiple possible conditions (e.g., poor, fair, good, and excellent). Concerning causal associations, bioassessments may support identification of general causation associated with broad environmental perturbations such as nutrient enrichment. Causal associations of a categorical type recognize individual impacts from multiple causative factors such as nutrient enrichment, habitat degradation, and toxic substances. Parameter-specific causal associations recognize the effects of specific stressors (e.g., low taxa richness and abundance of mayflies due to heavy-metal contamination). In the context of the protocol (Figure 1-2) presented here, the complexity of most urbanization impacts generally requires a study effort equivalent to a Level 3 or Level 4 (Barbour & Yoder, 2007) to effectively address the many variables involved. Alternatively, a Level 1 or 2 study might be appropriate in support of site selection (Step 3 of the protocol, Figure 1-2), given the ability of these studies to support general pass/fail determinations regarding attainment or nonattainment of broad ecological goals. Given the limitations of Level 1 and 2 studies, such efforts would need to be followed by more detailed studies at Level 3 or 4 in order to adequately develop linkages among biotic, habitat, and hydrologic and geomorphic parameters (Steps 4 and 5 of the study protocol, Figure 1-2).

Table 5-1. Degrees to Which Bioassessment Levels Support Key Water Quality Study and Management Questions (adapted from Barbour & Yoder, 2007).

	Condition Assessment		Causal Associations		
Data Condition Level	**Impairment/non-impairment**	**Multiple Condition**	**General**	**Categorical**	**Parameter Specific**
1	*	--	--	--	--
2	**	*	*	--	--
3	**	**	**	**	*
4	***	***	***	***	***

"--" indicates the data condition level does not support a particular study goal.

*** - Indicates fulfillment of condition assessment and categorical causal association study goals while minimizing Type 1 and Type 2 study errors.

** - Indicates fulfillment of condition assessment and general causal association study goals while minimizing Type 1 study errors (Type 2 errors not adequately addressed).

* - Indicates fulfillment of the indicated study goal; condition assessments minimize Type 1 study errors; no causal association ability.

Note: Type 1 and 2 study errors are (Type 1) suggesting impairment when it does not actually exist and (Type 2) failing to identify impairment because of the inability of the assessment tool to detect it.

The 12 critical elements of a viable bioassessment program, as adapted by Barbour and Yoder (2007) from U.S. EPA's 2002 Consolidated Assessment Listing Methodology (CALM), fall within three major program components of program design, study methods, and interpretation of assessment results (Figure 5-2):

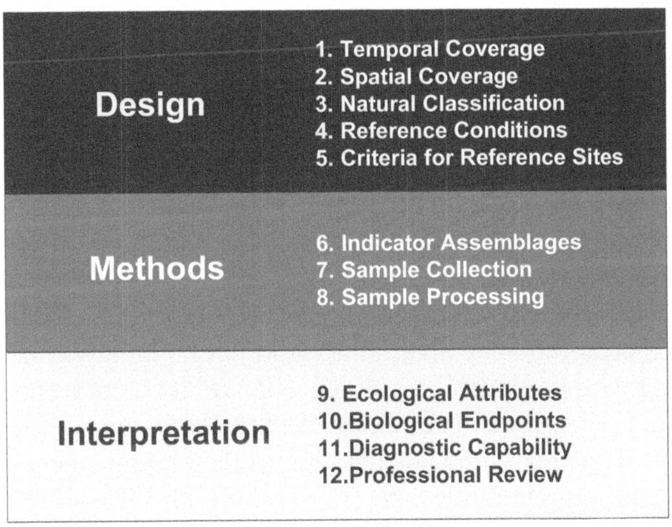

Figure 5-2. Critical Elements of Bioassessment Studies (adapted from Barbour & Yoder, 2007).

Design

1) **Temporal Coverage:** Temporal coverage for a given sampling program relates to the use of a consistent seasonal time frame for sampling (i.e., an "index period") that is a cost-effective alternative to year-round sampling. The optimal index period for sampling

should correspond to the reproduction, growth, emergence, and migration patterns of the organisms being studied.

2) **Spatial Coverage:** In order to support extrapolation of biological conditions beyond the exact sampling location(s), the following guidelines should be considered and incorporated into the program design. First, there should be base coverage in first- to third-order streams to provide data at a small enough scale to isolate important contributing stressors. This is due to the fact that the structure of ecological communities changes in a natural and predictable manner with increasing size of the ecosystem (in this case a stream, river, or entire watershed). Second, changes in land use along a stream or among multiple study streams should be considered. These protocols recommend selecting watersheds that represent the range of development and land use within the studied region. Third, the spatial density of sites and their placement with respect to upstream drainage area should also be considered. For example, Ohio's current sampling design for watershed assessment covers all streams in a watershed down to 1st or 2nd order, which provides a spatially even and dense coverage (Ohio EPA, 2005a). Data collected from these sites may be linked to regionally available data while establishing linkages among land use, hydrology, habitat and biota. Subsequent sampling may focus in more detail on identified impairments within particular watersheds.
3) **Natural Classification:** This program element relates to classifying waterbodies for the purpose of partitioning natural variability and distinguishing it from variability resulting from human-induced changes. Classification may be based on a combination of characteristics such as ecological regions, elevation, temperature (e.g., cold water versus warm water habitat streams), and other physical stream features for each waterbody type (e.g., large rivers, wadeable streams, headwater streams). The limitation of study streams for the purposes of these protocols generally limits considerations to wadeable streams. The selection of biological indicators, indices, and metrics is critical, and the investigator needs to understand that expectations for a particular indicator may vary with abiotic natural classification factors such as stream size or stream gradient. Most multimetric indices are calibrated based on factors such as drainage area. Expected species richness often changes with stream size, especially for fish; therefore, indicators might need to be scaled (e.g., expected number or percentage of taxa or species rather than absolute numbers) to account for these variables.
4) **Reference Conditions:** The regional reference site concept is intended to include the range of natural variability in biological condition due to natural differences in geology, climate, and so forth. Stoddard et al. (2006) propose some standardization to various definitions of reference condition that will depend on the level of "background" landscape disturbance that has occurred in an area. This effort goes a long way to reduce confusion related to describing various "types" of reference including natural (no human influence), historical condition, least-disturbed condition, minimally disturbed condition, and best-attainable condition. Together with consideration of the biological condition gradient concept of Davies and Jackson (2006), the process here for measuring responses to urban gradients is clearer.
5) **Criteria for Reference Sites:** There are a number of ways to select reference sites and the method will depend on the reference category selected (see Stoddard et al., 2006). One common method is to use nonbiological factors such as human population density and distribution, road density, and the proportion of land uses such as mining, logging, agriculture, urbanization, or other land uses as a primary basis for selecting reference

sites. The intent is to ideally select the best streams within a given study area. In light of the protocol, reference sites would be represented at the least-impacted range of sites along the urban development gradient (see the Site Selection section).

Methods

6) **Indicator Assemblages:** The term "assemblage" refers to an association of interacting populations of organisms in a given waterbody. For streams and rivers, biological indicators are generally separated into four main assemblages: fish, aquatic macroinvertebrates or "benthos," algae, and aquatic macrophytes or vascular plants. Including more than a single assemblage as part of study efforts adds to the confidence in observed results and can also enhance understanding about observed causes of impairment.
7) **Sample Collection:** Field methods should be standardized across sample locations to establish the validity and reliability of biological data generated by the study.
8) **Sample Processing:** Sample processing varies according to the assemblage being studied and can range from a cursory examination and documentation of presence or absence of a given organism or group of organisms to quantitative measurement of abundance and distribution among samples. In the more rigorous levels of sampling, the adequate number of organisms required for a valid sample is specified. The ultimate characterization of biological condition is dependent on the appropriateness and adequacy of sample processing.

Interpretation

9) **Ecological Attributes:** Development of biological indicators is dependent on the ability to measure ecological attributes of the aquatic assemblage being studied (Davies & Jackson, 2006). Ecological attributes are the aspects of an aquatic assemblage or community that correspond to the structure and function of the assemblage or community under a given biological condition. Ten primary ecological attributes have been identified as the basis for evaluating the biological condition gradient:

   - 1st – environmentally sensitive, long-lived, or endemic taxa;
   - 2nd – environmentally sensitive rare taxa;
   - 3rd – sensitive ubiquitous taxa;
   - 4th – taxa of intermediate tolerance;
   - 5th – environmentally tolerant taxa;
   - 6th – non-native taxa that tend to displace endemic taxa (i.e., invasive species);
   - 7th – refers to the health of individual organisms;
   - 8th – ecosystem function is large, comprehensive, and difficult to measure directly;
   - 9th – the spatial and temporal extent of observed detrimental impacts; and
   - 10th – the degree of connectivity exhibited by the ecosystem under consideration in the context of the larger watershed.

   A consistent approach to categorizing these attributes allows the biological condition of streams to integrate into a consistent framework for analyzing impairments.

10) **Biological Endpoints:** This program element relates to biological index development and identification of thresholds for determination of impairment. Index development can include single or multiple metrics (e.g., number of sensitive taxa, number of environmentally tolerant taxa) or statistical models predictive of the aquatic assemblage. Thresholds are the criteria describing the biological condition needed to support the ecological goals set in Step 1 of the protocol (see Figure 1-2).

11) **Diagnostic Capability:** The diagnostic capability of a program is directly tied to the extent to which the study effort characterizes the full gradient of human disturbance and associated biological response within the study area (Figure 1-2, Step 3 of the protocol).

12) **Professional Review:** Subjecting program study methods and assessment reports to rigorous peer review is the best way to ensure the generation of credible data.

The individual key program elements are described in further detail in Table 5-2 with regard to how data condition Levels 1 through 4 differ for each of the elements.

Ohio Example of Bioassessment Study Design for an Urbanizing Watershed

An intensive watershed survey is a standard way to obtain data to generate stressor and biological gradients in urban and urbanizing watersheds. Such studies at the state level are usually designed to address multiple goals including aquatic life management but can provide a solid baseline over which to quantify urban gradients. Ohio EPA uses what is called a geometric sample design to ensure a dense and spatially even distribution of monitoring sites typically across watersheds based on U.S. Geological Survey 11-digit Hydrologic Units (http://water.usgs.gov/GIS/huc.html) with a typical size of 125 mi^2 (Ohio EPA, 2005a). The Big Walnut Creek watershed near Columbus, Ohio, has historically had both urban and agricultural land uses; however, urbanization has greatly accelerated in the past 10 years. An Ohio EPA watershed survey documented impairments from development in the watershed (Ohio EPA, 2003), which in turn formed the basis for a TMDL determination (Ohio EPA 2005b) for Big Walnut Creek. The primary causes of impairment in this watershed were determined to be nutrient enrichment, pathogens, and habitat degradation, all of which are increasingly related to rapid urbanization.

The map shown for this example illustrates the spatial coverage of fish assemblage monitoring stations in the Big Walnut Creek watershed. Ohio's sampling distribution is dense but supports multiple environmental management goals. Comparison of biological assemblages in Big Walnut Creek, with historic data from other studies, documented losses of sensitive fish species associated in a subwatershed with the most rapid urbanization (Rocky Fork Big Walnut Creek) (Ohio EPA, 2003; Miltner et al. 2004). Short- term, follow-up monitoring in this watershed likely will focus on indicators to refine and understand the mechanism of the species loss (stressor identification process). Longer-term monitoring (~10 year cycle) will track the progress or decline from the implementation of management measures to control development related impacts from nutrients, pathogens, and habitat modification.

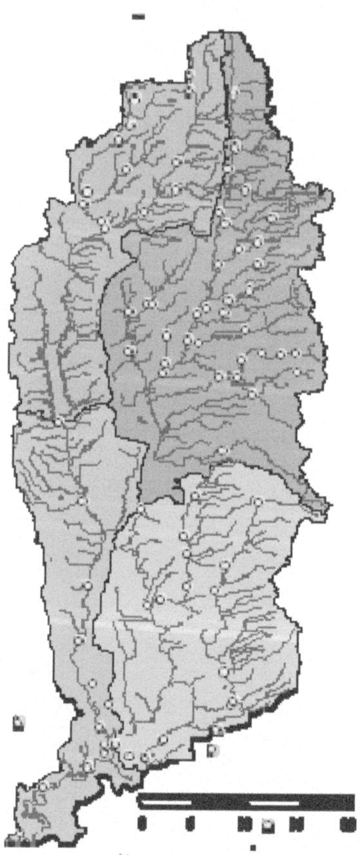

When studying wet weather and urbanization patterns, some emphasis should be placed on the second key program element, spatial coverage. To adequately address cumulative effects of urbanization, data are needed at a sufficient geographic scale to understand potential spatial limitations to impacts on the observed aquatic assemblages. For example, in instances where specific taxa appear to be missing, data of sufficient spatial coverage are needed to determine whether their absence is the result of a localized or more regional phenomenon. Cumulative effects of various stressors have been identified by numerous workers and are an integral part of understanding urban stressors (Booth et al., 2003). The spatial extent of impacts can exert a strong influence on biological performance in an area. Because of this, studies of wet weather and urbanization need to either be large enough to capture these scale influences or be able to access existing data to account for these limitations. This is another reason to first consider leveraging data, methodologies, or both from existing programs in an area that would provide a strong spatial basis for data interpretation. Studies with a narrow spatial context can yield data representing too restricted a gradient of conditions along the multiple spatial scales that influence biological assemblages. For example, studies at the stream reach level could lead to erroneous conclusions if the resulting data are not evaluated in the context of the larger surrounding watershed. Such sampling efforts might yield data conditions under which the investigator(s) cannot "see the forest for the trees." This could result in insufficient power to detect an impact when it becomes evident.

5.1.2 Identification of Existing Ecological Goals and Biological Measures

For the study protocols presented here, it is recommended that the investigators first explore existing goals that may be in place related to aquatic life uses in state water quality standards (WQS) for the water resource under consideration. A potential resource for this exploration is the U.S. EPA Web site (http://www.epa.gov/), which contains links to state WQS, including designated uses and antidegradation rules. Most states have their own Web sites that often contain a listing of waterbodies within the state or at least a narrative related to designated aquatic life uses for each waterbody. Researchers conducting such an assessment should work closely, as appropriate, with state management agencies to maximize available knowledge regarding the water quality of streams in the region. Using existing goals provides opportunities to build on the groundwork of prior efforts and, if well constructed, the goals will be explicitly linked to existing assessment methodologies. This substantially helps derive data quality objectives related to whether an ecological goal is being met for a given waterbody. For ecological goals, the burden of identifying attainment or nonattainment should be borne primarily by biological assessment methodologies. This approach allows the usage of habitat or other stressor data to be utilized for purposes of stressor identification. Although this difference in data usage may seem subtle, it allows for the direct measurement of biological parameters in lieu of intensive chemical analyses used to measure water quality and attainment or nonattainment of ecological goals.

A few states such as Ohio and Maine have derived biological criteria (biocriteria) that provide specific ecological endpoints for a given aquatic life use. For example, Ohio has specific indices including the IBI and MIwb for fish and the ICI for aquatic macroinvertebrates (Ohio EPA, 1987; 1989). These multimetric indices are specified in state rules for various streams based on tiered aquatic life uses, stream size, and ecoregion. Many other states use biological data to assess aquatic life uses but may or may not have specific guidelines for such assessments in their WQS standards. Where these methodologies do occur, they are often found associated with state water quality reports (CWA, Section 305(b)) and/or state TMDL assessments. These are typically available through state water quality agency Web sites. Alternatively, states often have assessment methodologies published on monitoring Web sites.

Table 5-2. Key Biological Assessment Components and Corresponding Levels of Rigor for Each Data Condition Level (adapted from Barbour & Yoder, 2007).

Key Bioassessment Program Elements	Data Condition Level 1	2	3	4
Design				
1. Temporal coverage	No index period is established and sampling is performed without regard to seasonal influences.	An index period is conceptually recognized, but sampling may take place outside this period for convenience or to be consistent with data generated from other existing programs.	Well-documented seasonal index period(s) is/are calibrated with data for reference conditions, but sampling may take place outside this period for convenience or to be consistent with data generated from other existing programs; data collected outside the index period are adjusted for seasonal influences.	Same as Level 3 but program administrative needs and index periods fully reconciled.
2. Spatial coverage	Individual site used for assessment of watershed condition; simple upstream/downstream and fixed station designs prevail.	Multiple sites used for watershed assessment; spatial coverage generally lacking for questions beyond general status or locally specific problem areas.	Spatial network of sample locations suitable for biological condition status assessments with single purpose design at a coarse scale (e.g., 8-digit USGS HUC units).	Comprehensive spatial network suitable for reliable watershed assessments in support of multiple water quality management programs at a more detailed scale (e.g., 11-14 digit HUC units).
3. Natural Classification	No partitioning of natural variability in ecosystems; does not incorporate differences in stream characteristics such as size or gradient.	Classification recognizes one stratum, usually geographical or fishery-based cold or warm water habitat; lacks other intra-regional strata such as watershed size, gradient, elevation, temperature, etc.	Classification based on a combination of landscape features and physical habitat structure; considers all relevant inter-regional characteristics and subcategories of specific stream types.	Fully partitioned and stratified classification scheme based on a regional approach.
4. Reference conditions	No reference condition normally established; assessment of impairment based on best professional judgment.	Reference condition based on biology of a "best" site or waterbody; regional reference sites not used.	Reference conditions based on site-specific data but used in watershed-scale assessments; regional reference sites recognized but few in number or without adequate coverage.	Established regional reference conditions for a number of sites within applicable waterbody types; adequately calibrated for assessing and monitoring biological condition within a given waterbody type.

Table 5-2. Key Biological Assessment Components and Corresponding Levels of Rigor for Each Data Condition Level (adapted from Barbour & Yoder, 2007).

Key Bioassessment Program Elements	Data Condition Level 1	2	3	4
5. Criteria for reference sites	No criteria except informal selection of control sites based on best professional judgment.	Basis is best professional judgment on the best biology in the best waterbody; minimal non-biological data used.	Nonbiological criteria used as basis but supported by narrative descriptors only; combines best professional judgment with narrative description of land use and site characteristics; chemical and physical data thresholds may be used as primary filters for site selection.	Quantitative descriptors used to support non-biological criteria; sites selected for characteristics believed to be supportive of best biological community condition; chemical and physical site characteristics used only as secondary and tertiary filters during site selection.
Methods				
6. Indicator assemblages	Gross observation of biota; single assemblage only; very low taxonomic resolution (e.g., order/family level for benthos; family level for fish).	Single assemblage (usually macroinvertebrates); low taxonomic resolution (e.g., family level) determined by experienced biologist(s).	Single assemblage with high taxonomic resolution ("lowest practical" or genus/species); if multiple assemblages, others at lower resolution or infrequently used.	Two or more assemblages with high taxonomic resolution ("lowest practical" or genus/species); can use each assemblage concurrently in bioassessments.
7. Sample collection	No formal documentation of methods; approach relies on operator skill and best professional judgment; no or minimal training.	Textbook methods used with no in-house adaptation; short courses (1-2 days) provided for new staff and periodically for all staff.	Methods periodically evaluated and refined as needed; detailed methods with good documentation; standard operating procedures (SOPs) periodically updated through in-house testing and development; formal quality assurance/quality control program in place; rigorous training for all applicable staff.	Same as Level 3 but includes multiple assemblages.
8. Sample processing	Biological samples processed in the field using visual guides; sorting and identification are dependent on operator skill and effort.	Organisms are identified and counted primarily in the field but processing is performed by trained staff.	Laboratory processing of all samples except for fish; a formal QA/QC program is in place.	Same as Level 3 but includes multiple assemblages.

Table 5-2. Key Biological Assessment Components and Corresponding Levels of Rigor for Each Data Condition Level (adapted from Barbour & Yoder, 2007).

Key Bioassessment Program Elements	Data Condition Level 1	2	3	4
Interpretation				
10. Biological endpoints	No formal index or assemblage-based endpoint; assessment may be based on presence or absence of targeted or key species; attainment thresholds not specified; coarse method that detects only high and low values.	A biological index or endpoint is established for specific water bodies, but is not likely calibrated to waterbody classes; index is likely limited to a single assemblage; regional reference conditions not quantified; presence/absence based on all taxa; thresholds based on best professional judgment regarding limited ecological attributes; limited to pass/fail determinations of threshold attainment.	A biological index or model is developed and calibrated for various classes of a waterbody type; the index pertains to a single assemblage; attainment thresholds based on well developed reference site data; can distinguish 3 to 4 increments along the biological condition gradient.	Biological index/indices or model(s) for two or more assemblages is/are developed; well developed regional reference condition data; able to detect status on a continuous scale along the biological condition gradient; power to detect at least 5-6 condition categories.
11. Diagnostic capability	No diagnostic capability due to low resolution information on taxonomy and ecological attributes.	Coarse indications of stressor-response relationships via low resolution ecological attributes (e.g., EPT macroinvertebrate taxa); little or no supporting analysis across spatial and temporal scales.	More detailed development of biological indicators to distinguish and support causal associations; usually involves refined taxonomy (at least genus level); supported by analysis of larger data sets and/or extensive case studies; developed for a single biological assemblage only.	Response patterns are most fully developed and supported by organized and extensive research and case studies across spatial and temporal scales; involves refined taxonomy; developed for at least two biological assemblages.
12. Professional review	No substantive review or inadequate reconciliation of comments and issues.	Internal review only with no review by a qualified and objective third party.	Outside peer review conducted, but with subjective selection of peer review.	Formal technical review process conducted, resulting in improvements to methods and reporting.

A summary of state monitoring programs and methodologies was produced by U.S. EPA (2002a), and links to these programs can be found on the U.S. EPA Web site (www.epa.gov). Table 5-3 presents a summary of the portion of total streams sampled in each state as well as the types of biota sampled (i.e., benthic macroinvertebrates or benthos, fish, and periphyton or algae and diatoms) as part of state bioassessment programs.

This document does not include a thorough review and assessment of each of the state bioassessment programs (Table 5-3) with regard to their associated overall data condition levels; however, the information related to the number of assemblages assessed and the associated level of taxonomy for sample processing provide insights into possible or probable data condition levels. As presented in Table 5-2, data condition Levels 1 and 2 generally have very coarse taxonomic resolution for a single assemblage, whereas Levels 3 and 4 are associated with one assemblage (Level 3) or multiple assemblages (Level 4) assessed at a relatively high taxonomic resolution. This information (Table 5-3) is intended to provide users of these protocols with an idea of how their intended bioassessment programs might fit into the larger context of available state bioassessment data. State and federal resources are available to assist with the development of project-specific DQOs for a desired data condition level. DQOs for state monitoring programs can provide a strong starting point for developing project-specific DQOs.

At the federal level, each U.S. EPA regional office also has a monitoring coordinator that could serve as a useful resource for this and other monitoring issues developed and implemented at local and/or regional levels. U.S. EPA also has a number of publications to help states develop biological assessment methodologies, although many states modify these or other similar methodologies to the ecosystems and waterbody types in their regions. The only nationally consistent methodology has been the Environmental Monitoring and Assessment Program (EMAP, Hughes et al., 2000). These methods (biological and detailed habitat assessments) have been used recently for the National Wadeable Stream Assessment (WSA) survey (Holdsworth et al., 2004; U.S. EPA, 2004; 2006). Methods are summarized in a number of recent publications. The U.S. EPA is performing comparisons among existing state methods and the National Wadeable Stream Assessment methodologies. The WSA methodology is among the most in depth and rigorous, especially with regard to habitat measures and would provide an advantage of leveraging existing data that is national in scope and availability. U.S. EPA also recently performed a comparison of a limited set of boating and wading methods used by federal and state agencies to assess the status of flowing waters that could serve as a valuable resource for assessment methods (Flotemersch et al., 2001).

Table 5-3. Summary of Stream Miles Covered, Biological Assemblages Assessed, and Habitat Assessment Types for State Bioassessment Programs (adapted from U.S. EPA, 2002a).

State Name	Total Stream Miles	Total Stream Miles Assessed	Assemblages Assessed						VB = visual based; QM = quantitative measurements; HY = hydrogeomorphology; O = other
			Benthos	Taxonomy	Fish	Taxonomy	Periphyton	Taxonomy	Habitat Assessment
Alabama	77,274	7,103.5	Y	Family, Genus	Y	Species	N	NA	VB
Alaska	>3 million	150 watersheds	Y	Genus	N	NA	UD	UD	VB, HY
Arizona	127,505	0	Y	Family, Genus, Species	N	NA	Y	Species (diatoms)	VB, QM, HY
Arkansas	87,617	245 stream segments	Y	Family, Genus, Species	Y	Species	N	NA	VB, QM, HY, O
California	211,513	unknown	Y	Genus, Species	N	NA	N	NA	VB
Colorado	107,403	n/a	Y	Genus, Species	Y	Species	UD	UD	VB, HY, O
Connecticut	5,830	961	Y	Species	Y	Species	Y	Species	VB
Delaware	2,506	2,506	Y	Genus	N	NA	N	NA	VB
District of Columbia	39	39	Y	Family	Y	Species	N	NA	HY
Florida	51,858	4,795	Y	Species	N	NA	Y	Species	VB
Georgia	70,150	1,416	Y	Genus	Y	Species	N	NA	VB, O
Hawaii	249	15	UD	UD	Y	Species	N	NA	VB, O
Idaho	96,200	16,742	Y	Species	Y	Species	Y	Species	VB, O
Illinois	86,021	15,304	Y	Family, Genus, Species	Y	Species	N	NA	VB, QM
Indiana	35,673	35,430	Y	Family	Y	Species	Y	Unspecified	VB
Iowa	71,665	2,018	Y	Family, Genus, Species	Y	Species	N	NA	VB, QM

Table 5-3. Summary of Stream Miles Covered, Biological Assemblages Assessed, and Habitat Assessment Types for State Bioassessment Programs (adapted from U.S. EPA, 2002a).

State Name	Total Stream Miles	Total Stream Miles Assessed	Assemblages Assessed						VB = visual based; QM = quantitative measurements; HY = hydrogeomorphology; O = other
			Benthos	Taxonomy	Fish	Taxonomy	Periphyton	Taxonomy	Habitat Assessment
Kansas	134,338	23,731	Y	Genus, Species	Y	Species	Y	Species (diatoms)	VB, QM
Kentucky	89,431	~30,000	Y	Family, Genus, Species	Y	Species	Y	Species	VB
Louisiana	66,294	–	Y	Family, Species	Y	Species	N	NA	VB
Maine	31,672	1,000	Y	Genus, Species	N	NA	Y	Genus, Species	VB
Maryland	17,000	6,142	Y	Genus	Y	Species	N	NA	VB, QM, O
Massachusetts	8,229	1,344	Y	Genus, Species	Y	Sub-species	Y	Genus (algae)	VB
Michigan	49,141	21,469	Y	Family, Genus	Y	Species	N	NA	VB
Minnesota	91,944	2,047	Y	Genus	Y	Species	N	NA	QM
Mississippi	84,003	1,365	Y	Genus	N	NA	N	NA	VB, O
Missouri	52,194	21,996	Y	Genus, Species	Y	Species	N	NA	VB, QM, O
Montana	176,750	9,076	Y	Family, Genus, Species	Y	Species	Y	Genus, Species	VB, QM, HY, O
Nebraska	81,573	16,314	Y	Genus, Species	Y	Species	N	NA	VB, QM
Nevada	143,578	602	Y	Family, Genus, Species	N	NA	UD	Genus (algae)	VB, QM, O
New Hampshire	10,881	400	Y	Genus	Y	Species	N	NA	VB
New Jersey	6,500	330	Y	Family	Y	Species	N	NA	VB
New Mexico	110,741	~5,875	Y	Family, Genus, Species	Y	Species	Y	Diatoms only	VB, HY, O

Table 5-3. Summary of Stream Miles Covered, Biological Assemblages Assessed, and Habitat Assessment Types for State Bioassessment Programs (adapted from U.S. EPA, 2002a).

State Name	Total Stream Miles	Total Stream Miles Assessed	Assemblages Assessed						VB = visual based; QM = quantitative measurements; HY = hydrogeomorphology; O = other
			Benthos	Taxonomy	Fish	Taxonomy	Periphyton	Taxonomy	Habitat Assessment
New York	52,337	16,000	Y	Genus, Species	Y	Species	Y	Species (diatoms)	QM
North Carolina	37,662	32,072	Y	Genus, Species	Y	Species, Sub-species	Y	Species (diatoms)	VB
North Dakota	54,427	14,426	Y	Family, Genus, Species	Y	Species	Y	Species (diatoms)	VB, HY
Ohio	29,113	9,535	Y	Family, Genus, Species	Y	Species	N	NA	VB
Oklahoma	78, 778	13,313	Y	Genus	Y	Species	N	NA	QM
Oregon	114,823	40,188	Y	Genus, Species	Y	Species	Y	All Algae?	QM
Pennsylvania	83,000	45,000	Y	Genus	Y	Species	N	NA	VB
Rhode Island	1,498	272.8	Y	Family, Genus, Species	N	NA	N	NA	VB
South Carolina	35,461	678.6	Y	Family, Genus, Species	N	NA	N	NA	VB
South Dakota	9,937	3.73	Y	Family, Genus, Species	N	NA	Y	Species	VB, QM, HY
Tennessee	60,187	24,233	Y	Genus	N	NA	N	NA	VB
Texas	191,228	266.9	Y	Family, Genus, Species	Y	Species	N	NA	QM
Utah	85,916	705	Y	Family, Genus, Species	N	NA	UD	UD	QM, O

Table 5-3. Summary of Stream Miles Covered, Biological Assemblages Assessed, and Habitat Assessment Types for State Bioassessment Programs (adapted from U.S. EPA, 2002a).

State Name	Total Stream Miles	Total Stream Miles Assessed	Assemblages Assessed						VB = visual based; QM = quantitative measurements; HY = hydrogeomorphology; O = other
			Benthos	Taxonomy	Fish	Taxonomy	Periphyton	Taxonomy	Habitat Assessment
Vermont	7,099	~800	Y	Genus, Species	Y	Species	Y	Unspecified	VB, HY, O
Virginia	50,329	15,540.4	Y	Family	N	NA	N	NA	VB
Washington	73,886	3,275	Y	Family, Genus, Species	Y	Species	Y	Genus	VB, QM, HY
West Virginia	32,278	5,745	Y	Family, Genus	Y	Species	N	NA	VB, QM, O
Wisconsin	55,000	24,422	Y	Family, Genus, Species	Y	Species	Y	Diatoms only	QM
Wyoming	113,422	2,639	Y	Genus, Species	N	NA	UD	UD	VB, QM, HY,

5.1.3 Selection of Biological Parameters

Level 4 bioassessment programs (Barbour & Yoder, 2007) are characterized by the use of multiple organism groups such as fish and aquatic macroinvertebrates. Inclusion of multiple organism groups in a monitoring program provides the most robust approach for assessing urbanization impacts. Different organism groups may have varying sensitivities to each of the stressors' association with urbanization, and the use of two groups minimizes Type 1 study errors in suggesting impairment when it does not actually exist, and Type 2 study errors in failing to identify impairment because of the inability of the assessment tool to detect it. Yoder (1998) and Barbour and Yoder (2007) discussed the importance of states using more than one organism group, especially where multiple stressors occur and in complex situations. The scenarios for which the sampling protocols are being developed in this project (urbanizing watersheds) are typically complex with multiple stressors often changing at the same time. These often include changes in flow regimes, habitat characteristics, sediment dynamics, temperature, and water chemistry, which can be affected by overland runoff, sanitary sewer overflows (SSOs) and combined sewer overflows (CSOs), direct discharges, and even local atmospheric deposits. Even the nature of dissolved materials changes with urbanization, including complex organic and particulate matter changing in type, concentration, and timing of delivery. Amid the vast number of potential environmental stressors, differential responses of organism groups and individual taxa greatly assist in isolating causative stressors related to urbanization.

Although integrated indices may serve as the determining criteria for attainment or nonattainment of the ecological goals (Figure 1-2, Step 1), individual metrics or indicators are widely used in stressor identification work and should be included as part of this protocol. Multimetric indices are based on a premise that single indicators are too narrow in their sensitivity to use as standalone indicators of aquatic life use attainment. Conversely, multiple individual metrics with differing

Minimum Recommendations for Biological Parameter Selection and Sampling

The following are provided as minimum recommendations for protocol users in selecting and sampling biological parameters:

- A bioassessment program targeted at a Level 1 or 2 data condition may be employed as part of the study site selection process. The data generated from such assessment efforts are useful for accepting or rejecting prospective study sites based on whether they are too severely degraded to resolve targeted stressor-response relationships. Data at this level of detail will not, however, be useful for establishing a biological condition gradient along the range of watershed development targeted by these protocols. Level 1 data, in particular, may be more important for use in public education and outreach campaigns (e.g., citizen monitoring efforts) than for consistent identification of biological conditions.
- A bioassessment program targeted at a Level 3 or 4 data condition must be used to support the higher level stressor diagnosis and management criteria development purposes of these protocols. This necessarily involves a higher level of taxonomic resolution (i.e., genus/species) for one (Level 3) or more (Level 4) biological assemblages.
- The state bioassessment program(s) relevant to the intended study area should be evaluated as a source for DQOs and sampling protocols, as well as to leverage existing data to provide a broader spatial and temporal coverage.
- If state bioassessment program(s) are either unavailable or inappropriate for use in the context of the intended study, protocol users should consider the U.S. EPA Wadeable Streams Assessment sampling and assessment methodologies as a starting point (Holdsworth et al. 2004, U.S. EPA 2004, 2006).

sensitivities generally provide a better overall estimate of ecological condition. Similarly, multivariate approaches generally summarize information provided by the response of multiple species or taxa. The decision on which individual biological parameters should be used as response indicators, in addition to integrated measures such as IBI and ICI, will depend on the expected urbanization related stressors.

Conceptual models such as the ones shown in Figures 5-3 and 5-4 should be constructed on the basis of the hypothetical response of species and species grouping (e.g., taxonomic, functional) to the physical, chemical, and flow regime stressors thought to be operating in a specific watershed or region of interest. For example, a suite of habitat sensitive species can be used or a group of organisms that has life history strategies that are sensitive to a stressor. An example for the fish community (Figure 5-3) is simple lithophilic spawning fish species (i.e., fish that spawn in streams with clean gravel beds) as an indicator for increased embeddedness, a condition in which fine sediments and silt fill voids in larger substrate particles in the stream bed, or increased stream bed material instability related to urbanization. Figure 5-4 offers an example for identifying stressors associated with rheophilic (fast-flowing water) benthic macroinvertebrates.

Figure 5-5 describes the general process for identifying potential ecological attributes that are useful in identifying environmental stressors (urbanization or development, in this case), as part of overall biological assessment efforts. These include community composition and community function attributes as well as other potential valued attributes where appropriate. Community indices generally function as tools to measure overall attainment or impairment of identified ecological goals (Figure 1-2, Step 1), and the metrics are used in helping identify causes (stressors) of impairment. Biological metrics for the fish community shown in Figure 5-5 were selected because of their anticipated sensitivity to direct or indirect impacts of hydrologic disturbance or physical habitat degradation from urbanization or development. WERF project 01-WSM-3 (Barbour et al.) provides an approach using macroinvertebrates as an example to identify biological response thresholds along one or more urban stressor gradients. This project successfully identified strong biological responses to stressor gradients using a similar approach on existing data from streams in Baltimore, Maryland; Cleveland, Ohio; and San Jose, California.

5.2 Habitat Sampling

Steps 3 and 4a of the protocol (Figure 1-2) presented here address the identification of relevant stream habitat parameters and collection of associated data. As with the methods for evaluating biological assemblages, a variety of habitat assessment tools are in use around the world that range from very qualitative to quantitative (see Bain & Stevenson, 1999; Kaufmann, 1993; Kaufmann et al., 1999; Rankin, 1995; Barbour et al., 1999). In general, the two types of stream habitat assessment methods are the more qualitative rapid bioassessment protocols, such as U.S. EPA Rapid Bioassessment Protocols (RBPs; Kaufmann, 1993) and Ohio EPA Qualitative Habitat Evaluation Index (QHEI, Rankin, 1989) as compared to more quantitative and measurement intensive protocols such as the U.S. EPA EMAP (Kaufmann et al., 1999) or WSA approach. U.S. EPA, through an aspect of their WSA study, is currently performing comparisons among various state methods and the WSA approach. This work may help identify attributes where the WSA or similar quantitative methods provide important precision over the rapid methods needed to meet the DQOs of an urbanization study.

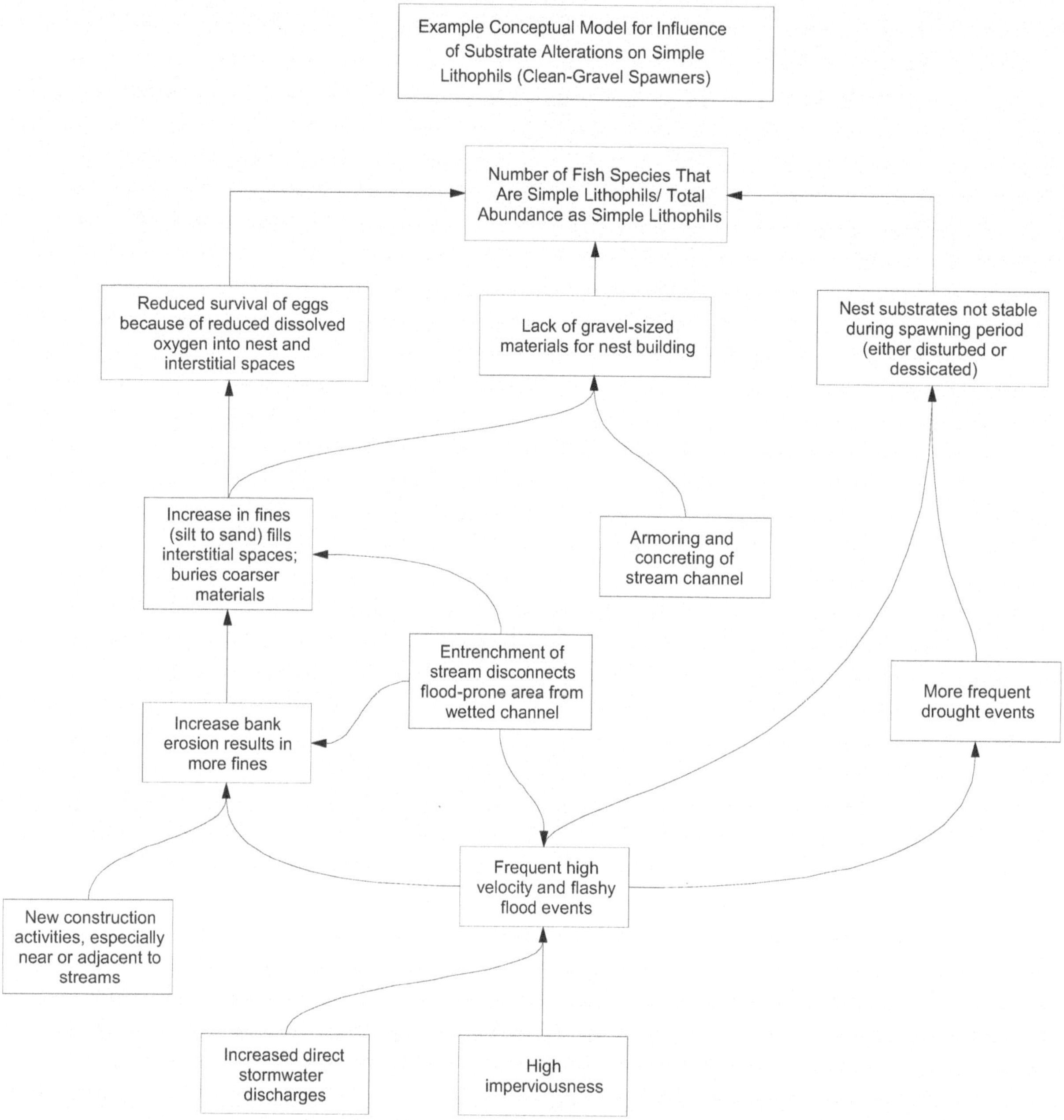

Figure 5-3. Stressor Identification Example for Lithophilic Spawning Fish.

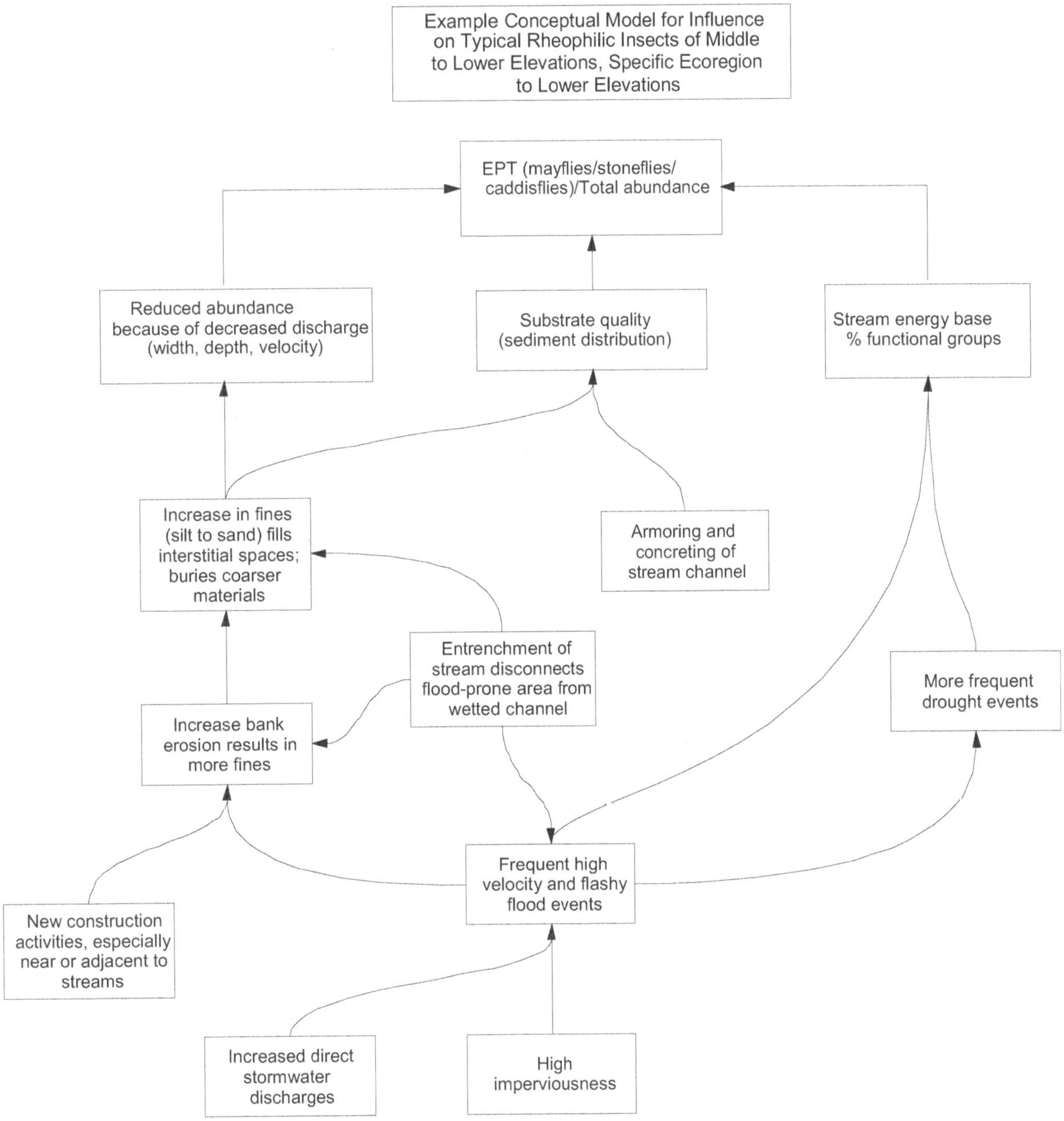

Figure 5-4. Stressor Identification Example for Rheophilic Macroinvertebrate Insects.

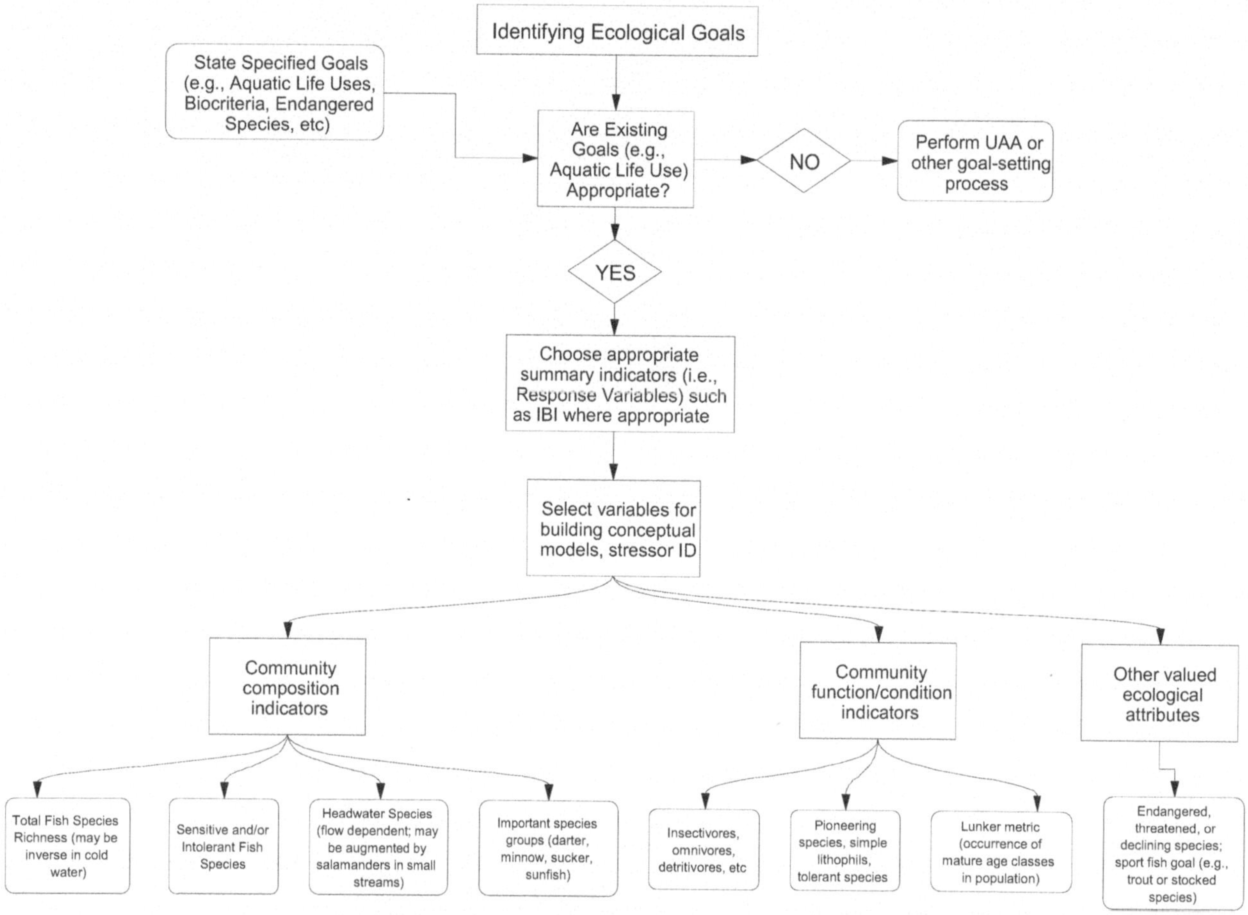

Figure 5-5. Protocol for Identifying Relevant Biological Study Measures (Example Depicts Selection of Study Metrics for the Fish Community).

At a minimum, we recommend that existing tools be examined first to see if they will provide sufficient resolution and breadth to identify important habitat limitations. Figure 5-6 summarizes habitat aspects that should be considered for this process. Physical habitat measures identified in Figure 5-6 were selected in light of their sensitivity to development-related alteration, whether direct (e.g., encroachment on riparian zones) or indirect, such as flow regime alteration impacts on channel characteristics.

The physical template of streams has long known to be important in determining the type and condition of the ecological fauna present in a stream (Hynes, 1970). It is essential to be able to assess habitat conditions to determine what a natural stream of a given type should look like in the absence of human alteration (i.e., physical habitat integrity) (Petts, 2000; Rabeni, 2000). Although a wide range of habitat tools have been used by various investigators to assess habitat, the precise tools selected will vary with the geographic setting and the scope of the study under consideration. Ideally, investigators will have access to a toolbox of methods from which they may choose. Most urban studies addressed by the protocol should have two major components: 1) a baseline effort to measure biological and physical aspects of the ecosystem in relation to biocriteria or other reference-based ecological goals, and 2) a stressor-identification phase during which impaired waters and initially identified stressors are examined in more detail to understand the mechanism of impairment and to build a toolbox of protection/restoration options in the form of management criteria. The first phase should be spatially detailed at a watershed scale and, as discussed under the bioassessment section, could rely on cooperation with existing efforts (e.g., state watershed surveys). The second phase may be more spatially limited but may be more detailed in terms of precision.

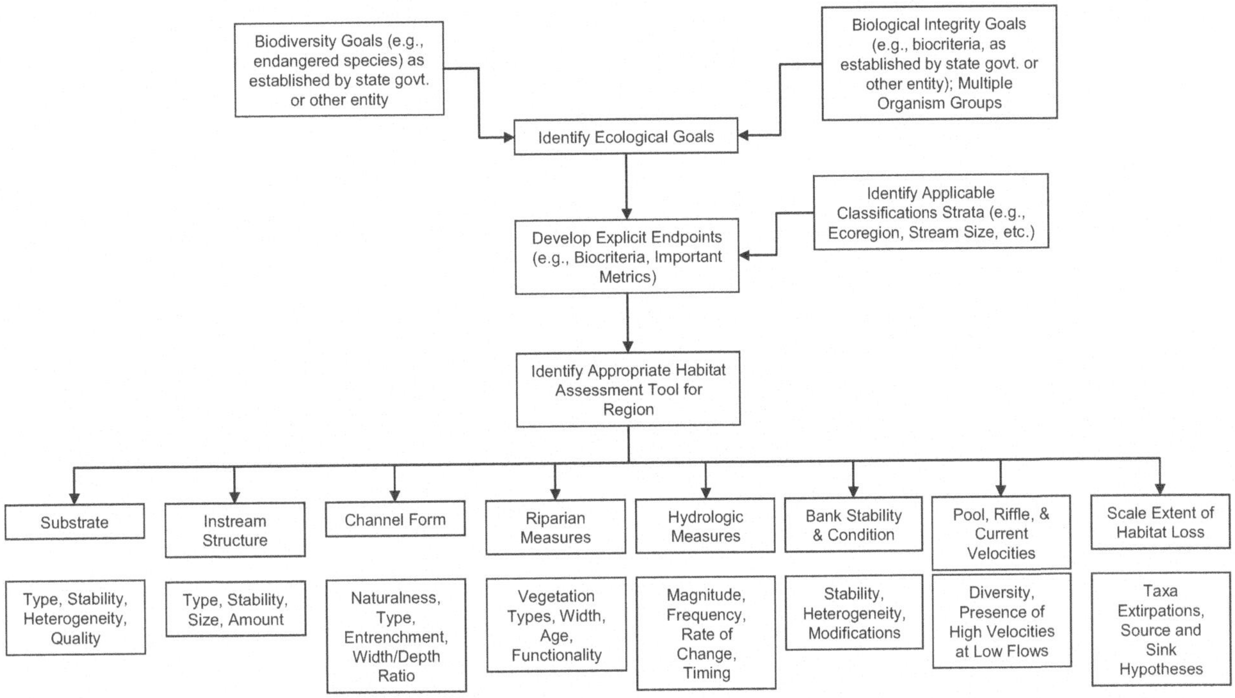

Figure 5-6. Protocol for Identifying Relevant Physical Habitat Study Measures.

To understand patterns at the scale of a watershed and larger (e.g., ecoregion) qualitative but robust and tested methods such as EPA's RBP habitat assessment tool (Plafkin et al., 1989; Barbour et al., 1999), the Maryland Physical Habitat Index (PHI, Paul et al., 2002), the River Habitat Survey (RHS) of the United Kingdom and Australia (Parsons et al., 2002; Davenport et al., 2004), AUSRIVAS in Australia (Parsons et al., 2002) or the QHEI (Rankin, 1989) have provided solid baseline information that has been associated with the biological indices and biological assessment methods outlined here. Another useful resource is the survey of state habitat evaluation methods in use as of the mid-1990s (Bain & Stevenson, 1999) and the Platts et al. (1983) summary. Armentrout (1998) provides a glossary of standard habitat terminology that is helpful. Scale influences are important to understanding habitat dynamics (Petts, 2000; Habersack, 2000) and it would be very difficult to collect sufficient data to understand these issues with local, case history collection approaches only (Bartleya & Rutherfurd, 2005).

Some of the data described above may still be suitable for the Phase II stressor identification process; however, there are more quantitative approaches such as the U.S. EPA EMAP habitat approach (Kaufmann et al., 1999; Hughes et al., 2000), USGS methods (Fitzpatrick et al., 1998) and the detailed approaches used in habitat suitability indices (HSI) typically for instream flow modeling (Bovee et al., 1998; Stalnaker, 1994). Booker and Dunbar (2004), for example, explored the HSI approach to urban streams in the United Kingdom. Where possible, these approaches should be considered within a geomorphic framework (e.g., Chessman et al., 2006), many of which are compatible with these habitat assessment approaches.

These more detailed tools can also be leveraged with data collected by state, federal, and local agencies. The U.S. EPA national wadeable stream study has recently finished a nationwide-scale data collection effort that provides data across the United States at randomly selected wadeable stream sites. USGS has data available from most of its NAWQA sites that are not randomly distributed but also some intensive data collection efforts in urban areas.

A comparison of the habitat assessment methods used around the world to assess streams finds many similarities in the attributes measured even if the mode of measurement differs (see Parsons et al., 2002). Relatively quick methods such as the EPA RBP method or QHEI in combination with the biological data collected should allow investigators to identify impaired waters and gain insight into whether habitat is a limiting factor and which habitat factors might be most important (Phase I). At that point, a stressor identification approach can be initiated to isolate key responsible stressors (U.S. EPA, 2000). One approach used for urban streams is a layered hierarchical approach outlined by Novotny et al. (1997) that incorporates habitat stressors at several levels of the model. Subsequent sampling may then be more detailed in order to provide a more specific and mechanistic explanation of the limiting factors (Phase II). These approaches may collectively be referred to as "active adaptive management" (Allan & Curtis, 2005) in which monitoring information is an integral part of stream management. In the following sections some specific habitat features are suggested for inclusion in any habitat assessment (Phase I and II), particularly in urbanizing waters.

An examination of habitat indices in use identifies a number of common habitat features across methods. Some of these will form the basis of a Phase I study but can also be used in Phase II. Other approaches may be too costly to apply throughout a watershed, but can be linked to Phase I approaches and applied selectively to a specific case study. Provided here are a brief discussion of these methods and some important references. Again, some of these approaches might have regional limitations due to their development in and applicability to specific geographic areas. As with the biological data, it is important to have sampled or have access to reference sites for a region to understand how habitat conditions may vary across the range and type of reference sites available (e.g., pristine *v.* least impacted *v.* best attainable).

5.2.1 Stream and River Substrates

All habitat indices and/or evaluation methods include one or more measures of stream substrate sizes and condition. Biological assemblages and individual species are strongly associated with substrate size and condition (Waters, 1995; Berry et al., 2003). Qualitative approaches typically use visual examination to quantify predominant sizes of materials and to assess common measures of condition such as embeddedness or silt cover. Numerous quantitative approaches have been developed including pebble counts (Wolman, 1954; Bunte & Abt, 2001) on riffles or across a stream reach (zigzag method of Bevenger & King, 1995). Some recent approaches that have been well correlated with biological indicators include the riffle stability index of Kappesser (2002) and the similar relative bed stability metric extracted from the detailed EMAP habitat assessment methodology (Kaufmann et al., 1999). The observed effects of urbanization on flow regime, shear stress, and bank erosion in streams are associated with changes in stream bottom substrate characteristics. These development-induced changes in stream substrates are likely important mechanisms influencing biotic assemblages in urbanizing watersheds. Data to quantify changes in substrate dynamics must be of sufficient resolution to detect differences across the scale of flow regime gradients being measured. More detailed geomorphic assessments may be needed to fully understand the dynamics of bedload transport that are absent from most habitat assessment methods (Muhar & Jungwirth, 1998). Although such methods as QHEI may be useful to identify four or five categories of degraded substrate, more subtle changes that may be of substantial importance in understanding cumulative impacts may require the more detailed measures mentioned here. The U.S. EPA EMAP habitat methodology (Kaufmann et al., 1999) provides a useful section that quantifies the precision of its metrics. Such an approach should be more widely adopted and can be used to help select the proper tool given the stage of the study being undertaken.

5.2.2 Instream Structure

As with substrate, a wide range of methods is used to quantify instream structure, also termed instream cover. Instream structure refers to those physical features of the stream that provide escape areas, such as logs and woody debris, slack areas for rest, winter refuges (undercut banks, rootwads), or other similar functions. More qualitative methods provide quick visual estimates of the amount and quality of cover (QHEI, RBP), whereas more quantitative methods (see Bain & Stevenson, 1999; Kaufmann et al., 1999) tend to use transects and estimate the number of various of types and sizes of materials (e.g., large woody debris). With urbanization, estimates of quality of instream structure are important because flashy flow regimes can act to isolate structure out of the low-flow channel because of excessive channel widening resulting from more frequent storm events and subsequent reductions in flows during dry periods.

5.2.3 Channel Condition and Bank Erosion

The manner in which various habitat methods assess channel condition varies greatly. The importance of natural channel form is underscored by its central place in many geomorphic characterization approaches (e.g., Simon & Hupp, 1986; Rosgen, 1994; Montgomery & Buffington, 1997; Hey, 1997; Petts, 2000). Most of the habitat assessment tools focus on the condition and morphology of channel form and ratings of modifications compared to natural. More detailed tools (e.g., EPA EMAP) rely on detailed mapping of a stream profile, which can be used to extract metrics related to channel condition (Kaufmann et al., 1999). Recent geomorphic approaches have focused on identifying the "stable" stream channel and measuring departure from this reference (e.g., Rosgen, 1999).

Most habitat assessment tools include one or more measures of bank condition such as the visual streambank soil alteration rating of Platts et al. (1983). More detailed bank assessment tools also exist such as ways to measure loss of bank materials (Pfankuch, 1978 in Rosgen, 1996; bank erosion hazard index of Rosgen, 2001).

5.2.4 Riparian Zones

The nature of the riparian zones of streams has received perhaps the most attention of any stream habitat component because of their importance as natural filters, shading, and their central importance to the natural food web of streams. It is clear, however, that preservation of natural riparian conditions alone is insufficient to protect and/or restore streams. Riparian zone widths and quality are also habitat attributes common to all habitat quality assessment methods. More qualitative methods (RBP, QHEI) tend to focus on major functions related to shading or width of the riparian zone with less information quantified on structure, function, or typology of particular riparian zones.

The EMAP and NAWQA habitat assessment protocols in general provide more detailed and precise assessments of riparian characteristics compared to simpler, visual assessments. The EMAP protocols, for example, include repeated densitometer measures as well as vegetation structure and type in the canopy, mid-layer, and understory (Kaufmann et al., 1999). The NAWQA protocols also include transect-based densitometer measures (Fitzpatrick et al., 1998). Compared to the RBP visual method, Kaufmann found that these more quantitative methods have less variation and higher precision (Kaufmann et al., 1999). Winward (2000) details three methods for monitoring the vegetation and associated characteristics of riparian areas. These methods also address some measurement of bank condition covered below. The DQOs established for a study will determine the necessary level of accuracy and precision and support the selection of an appropriate riparian assessment tool.

5.2.5 Streamflow/Current Velocity

The fact that flow often has a profound influence on aquatic life is well known (Poff et al., 1997), although it is more difficult to measure than other habitat features due to the strong temporal nature of its influence. A large part of stream management, especially in the arid Western United States, is focused on identifying the limiting effects of flow; a well-developed approach exists for such waters (IFIM, Stalnaker, 1994) especially related to salmonids and minimum flow needs. Petts (2000) summarized hydrological influences on rivers as follows:

Rivers, however, are characterized by strong hydrological and geomorphological gradients, which relate to highly dynamic processes of three types: hydrological processes (e.g., flooding, desiccation, surface-water and groundwater interactions), geomorphological processes (e.g., channel bed degradation and aggradation, bank erosion, floodplain accretion, channel avulsion), and hydraulic processes (including high shear stresses during spates and stagnation in dead zones during low flows) (Petts, 2000).

Most habitat indices are taken at single points in time and depend on indirect measures of the effects of flows such as eroding banks or degraded structure, or identification of current types available during typical low-flow sampling periods. Many such measures are often correlated with aquatic life; however, for understanding the influence of altered hydrology as urbanization proceeds, we suggest more detail may be needed. The use of various hydrological modeling approaches, verified with instream flow measures (gages) is important (Barbour et al., in press). Richter and others (1996) and Baker et al. (2004) provide various flow statistics that can be derived from USGS gages. Gages do not exist for most small headwater streams suggested for sampling in this protocol, so temporary gages may need to be installed.

Minimum Recommendations for Habitat Parameter Selection and Evaluation

The resolution needed in the habitat endpoints (DQOs) drives the choice of habitat tools needed for a particular study. The following are provided as minimum recommendations for protocol users in selecting and evaluating habitat parameters and have been successfully used to explain four to six levels of ecological condition. If an end point requires more precision (e.g., significant changes in bed particle size), more precise tools or approaches may be needed (e.g., Kaufmann et al.,1999):

- The state bioassessment program(s) relevant to the intended study area should be evaluated as a source for habitat evaluation protocols as well as to leverage existing data to provide a broader spatial and temporal coverage for habitat conditions.
- If state bioassessment program(s) are either unavailable or inappropriate for use in the context of the intended study, protocol users should consider the U.S. EPA RBP (Plafkin et al., 1989) or the Ohio EPA QHEI (Rankin, 1989) methodologies as a starting point

It is difficult to be explicit in prescribing specific techniques for monitoring habitat conditions in urbanizing streams. The myriad of currently available habitat evaluation techniques primarily results from several key factors:

1. The limited resources often available to managers of natural resources for habitat issues relative to study and management of stressors such as toxicants;
2. The penchant within some bureaucracies for developing "original" methods; and
3. The justifiable need to adjust tools to account for geographic differences across ecoregions and stream types.

Leveraging among regional users of habitat and geomorphic data can provide useful information that cannot be extracted from case studies alone. Considering a sufficiently broad urban gradient of habitat conditions reduces the risk of misidentifying the extent of habitat stressor impacts on aquatic life. For the broad-scale studies in Phase I, sampling design should follow what was laid out under the bioassessment section. Phase II work may well be more site-specific, but the study design must be such that urbanization effects should be readily measurable.

5.3 Screening Acutely Impaired Sites

The primary intent of water chemistry sampling protocols presented here is to support the selection process for study sites. The key concern for water chemistry related to site selection is to avoid the inclusion of acutely impaired sites. These are sites that exhibit water quality that is sufficiently poor to potentially obscure the effect of other development-related stressors. Primary non-chemical stressors related to urbanization include changes in hydrology and degradation of aquatic and riparian habitat. Table 5-4 provides a list of minimum recommended water quality parameters to consider during screening of prospective stream sampling locations. Water quality standards, often specific to designated uses for streams, are generally available from state governments for the parameters listed in Table 5-4. Sample results may be compared to established WQS for each parameter for an initial screening of prospective sites. A more relevant application of such data, however, is the comparison of observed values among the prospective study sites relative to one another. Sites that exhibit high concentrations of one or more of the recommended parameters, either in comparison to WQS or relative to other sites, could potentially be exhibiting evidence of development-related impacts.

Table 5-4. Summary of Minimum Recommended Water Quality Parameters for Study Site Screening.

Analyte	Volume Required	Method	Minimum Detection Limit	Units	Holding Time
Temperature	In field	Probe	NA	°C	NA
Dissolved Oxygen (DO)	In field	Probe	NA	mg/L	NA
Turbidity	In field	Probe	NA	NTU	NA
Suspended Sediment Concentration (SSC)	0.2 to 1 L	ASTM D 3977-97	NA	mg/L	NA
Escherichia coli (*E. coli*)	Not specified	ASTM D 5392-93	NA	*E. coli*	6 hrs

Notes:
NA = not applicable.
NTU = nephelometric turbidity units.

Elevated surface water temperatures can result from thermal pollution generated by permitted and illicit discharges. Elevated water temperatures can also be the result of increased penetration of sunlight into the stream channel due to destruction/removal of stream corridor vegetation (Horne & Goldman, 1994). Low dissolved oxygen levels can result from pollutant discharges and associated increases in either chemical or biological oxygen demand. Elevated turbidity and suspended sediment concentrations may be indicative of imported sediment resulting from upstream grading/earth moving activities or accelerated instream erosion due to altered urban hydrology. Observations related to these erosion indicators should be coordinated with habitat analysis of stream substrate conditions to further identify potential erosion impacts. *E. coli* contamination suggests contamination associated with human or animal waste, typically from inappropriate cattle grazing practices, improper pet waste disposal, or failing septic systems.

Such sites should be scrutinized prior to inclusion as part of a sampling program. Some states have categorized sediment and water chemistry values in relation to reference conditions or based on the potential to impair or affect aquatic life. Kelly and Hite (1984) characterized extreme values of sediment metal for Illinois streams, and others have done similar analyses that can be used to censor sites with heavily contaminated sediment. Burton (2002) reviewed the use of sediment values around the world. Certain biological signatures (sites with high prevalence of fish deformities, eroded fins, lesions, and tumors) are associated with toxic chemical stressors and could also be used to identify and remove acutely impaired sites (Yoder & Rankin, 1995a, b). Also, since the protocol is intended to evaluate the wet weather impacts of stormwater apart from water quality, watersheds and associated streams with CSOs, SSOs, and other significant point source discharges should be excluded from consideration as study sites (Yoder et al., 2000).

CHAPTER 6.0

Summary, Conclusions, and Recommendations

6.1 Summary

The objective of this research project was to develop a protocol that stormwater managers can use to evaluate how alternative land use development patterns and runoff controls impact the ecology of receiving streams in urbanizing watersheds. A review of recent literature led to the conclusion that this evaluation could be accomplished through the linkage of hydrologic metrics like the $T_{0.5}$ to indicators of biological health such as EPT richness or the B-IBI.

A pilot study of eight watersheds with a gradient of urban development in the North Carolina Piedmont was conducted and demonstrated that the $T_{0.5}$ hydrologic metric is responsive to changes in land use and to alternative runoff control scenarios. It was also found that the ecological health of streams in the North Carolina Piedmont, as measured by macroinvertebrate indices, is responsive to the $T_{0.5}$ metric, a finding similar to that found by Booth et al. (2004) in his study of the streams in the Pacific Northwest, a region of completely different hydrology. Furthermore, it was demonstrated that limited flow monitoring and macroinvertebrate data could be used in conjunction with hydrologic and hydraulic modeling (through EPA SWMM) to estimate how the biotic integrity of streams in a developing watershed might be affected through changes in land use patterns and runoff control scenarios.

Macroinvertebrate response was found to also be related to bed substrate size. An erosion potential index was shown to be valuable in demonstrating the geomorphic impacts of alternative stormwater management plans and runoff control strategies. Lastly, the importance of stream channel erosion control was highlighted in the finding that stream channel geometry influences hydrologic metric values through its impact on conveyance capabilities and loss of flood attenuation due to disconnectivity from floodplains.

6.2 Conclusions

The following conclusions were drawn from this project:

- Hydrologic metrics, which reflect the change in the hydrologic regime that occurs with urban development, have a higher correlation to the biologic health of a stream, as measured by macroinvertebrate indices, than indirect measures of urbanization such as percent imperviousness.
- Of the nine hydrologic metrics examined, the $T_{0.5}$ is the most responsive to changes in land use and mitigation through runoff controls.

- Physically based mathematical models that simulate runoff and stream flow (like SWMM5) can be used to estimate the value of the hydrologic metric T0.5 for alternative land use patterns and/or runoff control strategies in developing watersheds. This was done by first determining T0.5 values for watersheds of varying levels of urbanization in the same region and comparing those T0.5 values with those of urbanization watersheds. This allows stormwater managers to infer how alternative stormwater management plans and strategies will affect the biologic health of the receiving stream.
- Minimization of channel erosion is important because channel geometry directly influences hydrologic metric values because of its impact on conveyance capabilities and loss of flood attenuation due to disconnectivity from floodplains.
- The geomorphic impact of alternative stormwater management plans and/or runoff control strategies can be evaluated using computer simulations models (e.g., SWMM5) by using the stream channel erosion potential index.
- Research efforts to determine the biologic health of streams in study watersheds must reflect a data condition level of 3 or 4, generally associated with an intensive collection effort. A higher data condition level is required due to the inherent complexity of urbanization impacts on stream ecosystems. The study should optimally use multiple organism groups, such as fish and aquatic macroinvertebrates, in order to provide a more robust approach for assessing urbanization impacts. Also, multiple individual metrics of the biological community would optimally be used, with each reflecting different environmental sensitivities. Utilizing multiple community metrics with differing sensitivities to environmental stressors provides a better reflection of the overall biotic integrity of the stream(s).
- Concerning evaluation of habitat, a two-phase approach may be implemented. Phase I would utilize a quick or rapid assessment approach, such as the EPA RBP or the Ohio EPA's QHEI index, whereas Phase II would use a more intensive approach to refine the analysis of causes of degradation.
- Ecological monitoring data for biological and habitat parameters are essential to implementation of the protocol. The availability of data that meets DQOs for this type of work may not be readily available in all parts of the country, although state agencies and the U.S. EPA are working to refine monitoring programs and develop tiered aquatic-life uses. The choice of biological indicators and end points can have important consequences on study results, such as the example used here. Results from another recent WERF project (Barbour et al., in press) are particularly relevant for parts of the country in which adequate biological monitoring data are scarce or absent. The project provides an approach for generating biological indicators that are sensitive to the suite of urban stressors specific to a region or study area.
- In small communities that cannot afford to conduct the data sampling and computer modeling required by the protocol, the following minimum set of runoff control criteria for new development should be applied to provide improved protection of streams from a hydrologic, geomorphic, and biologic perspective:
- The peak discharge from the 100-, 10- and 2-year storm shall not exceed the peak discharges for these storms in the pre-development condition.

- An extended detention pond with at least a 40-hour drawdown time that is sized according to the WEF/ASCE (1998) Manual of Practice on Urban Stormwater Management shall be required on new development. If local standards requiring longer draw-down times are in place, these should be implemented.

6.3 Recommendations

It should be borne in mind that the watersheds in the North Carolina Piedmont on which data were gathered do not have properly sized BMPs or runoff controls for small storms; nor do the watersheds reviewed in the literature as far as is known. This research assumed that the relationship of the macroinvertebrate indices to the $T_{0.5}$ is sensitive only to the absolute value of the $T_{0.5}$ and not to other hydrologic effects that may not be captured by the $T_{0.5}$ such as the protracted duration of flows smaller than the 0.5-year storm that result when extended detention storage with 40-hr drawdown is applied to a watershed. It should be possible to indirectly assess the effect of protracted low flows using the erosion potential index because it is a measure of the geomorphic stability of the channel and, hence, the integrity of the benthic community. But this is engineering judgment, not proven fact. Therefore, it is recommended that a carefully conducted experiment be conducted on an urbanizing watershed of 1–3 square miles, with runoff controls and BMPs designed according to the criteria recommended in this study (see Conclusions above) to relate macroinvertebrate health to $T_{0.5}$ as the watershed develops and compare this data to watersheds that are in various states of urbanization but no significant number of runoff controls. Care must be taken in this experiment to control runoff during the construction phase or the macroinvertebrate indices will reflect the construction impacts rather than the built environment impacts.

It is believed that implementation of stormwater management practices that reduce the overall volume of runoff through infiltration or evapotranspiration will also aid in moving towards a more natural hydrologic flow regime that will allow for ecologically healthy receiving waters. The impact of such practices depends upon the extent to which they are implemented. Future research is needed to quantify the biotic impacts of volume reduction through practices such low impact development.

Water quality was not addressed in this work. From the standpoint of protecting the biotic integrity of receiving streams in urbanizing watersheds, the researches in this study believe that water quality is not the main issue; the runoff controls required to achieve hydrologic and geomorphic stability must include BMPs, which if properly designed, will remove pollutants from stormwater runoff. Nevertheless, the federal Clean Water Act is based on water quality criteria in receiving streams, and stormwater regulations now require that TMDL calculations include stormwater runoff. Therefore, it is recommended that future research build on this protocol, by adding water quality sampling of constituents identified in the water quality criteria applicable to the receiving stream, adding water quality to stormwater models as a runoff parameter, and simulating the removal of pollutants by runoff control practices. Currently available simulation models, including EPA's SWMM5 model, are a long way from being able to simulate credibly either runoff quality or constituent removal in BMPs.

APPENDIX A

Table A-1. Summary of Recommended Hydrologic Indices That Best Represent the Biologically Critical Portions of the Flow Regime, by Stream Type. (After Olden and Poff, 2003).

		Definition	Intermittent		Perennial				
			Harsh Intermittent	Intermittent Flashy or Runoff	Snowmelt	Snow & Rain	Superstable or Stable Groundwater	Perennial Flashy or Runoff	All Streams
Magnitude of Flow Events: Average Flow Conditions									
M_A3	Variability in daily flows 1	Coefficient of variation in daily flows				x	x		x
M_A5	Skewness in daily flows	Mean daily flows divided by median daily flows							x
M_A8	Ranges in daily flows	Ratio of 10th/90th, 20th/80th and 25th/75th percentiles in daily flows over all years					x		
M_A9	Spreads in daily flows	Ranges in daily flows (MA6–8) divided by median daily flows		x					
M_A10	Spreads in daily flows	Ranges in daily flows (MA6–8) divided by median daily flows						x	
M_A11	Spreads in daily flows	Ranges in daily flows (MA6–8) divided by median daily flows							x
M_A16	Mean monthly flows	Mean monthly flow for all months	x						
M_A18	Mean monthly flows	Mean monthly flow for all months		x					
M_A21	Mean monthly flows	Mean monthly flow for all months		x					
M_A22	Mean monthly flows	Mean monthly flow for all months	x						
M_A26	Variability in monthly flows	Coefficient of variation in monthly flows for all months						x	
M_A29	Variability in monthly flows	Coefficient of variation in monthly flows for all months			x				
M_A34	Variability in monthly flows	Coefficient of variation in monthly flows for all months	x						
M_A37	Variability across monthly flows 1	Variability in monthly flows divided by median monthly flows, where variability is calculated as range, interquartile, and 90th–10th percentile.		x					
M_A40	Skewness in monthly flows	(Mean monthly flow—median monthly flow)/median monthly flow			x				
M_A41	Mean annual runoff	Mean annual flow divided by catchment area					x	x	x

		Definition	Intermittent			Perennial			
			Harsh Intermittent	Intermittent Flashy or Runoff	Snowmelt	Snow & Rain	Superstable or Stable Groundwater	Perennial Flashy or Runoff	All Streams
$M_{A}44$	Variability across annual flows	Variability in annual flows divided by median annual flows, where variability is calculated as range, interquartile, and 90th–10th percentile.				x			
Magnitude of Flow Events: Low Flow Conditions									
$M_{L}1$	Mean minimum monthly flows	Mean minimum monthly flow for all months	x						
$M_{L}4$	Mean minimum monthly flows	Mean minimum monthly flow for all months							x
$M_{L}6$	Mean minimum monthly flows	Mean minimum monthly flow for all months		x					
$M_{L}13$	Variability across minimum monthly flows	Coefficient of variation in minimum monthly flows	x		x	x			
$M_{L}14$	Mean of annual minimum flows	Mean of the lowest annual daily flow divided by median annual daily flow averaged across all years				x	x	x	
$M_{L}15$	Low flow index	Mean of the lowest annual daily flow divided by mean annual daily flow averaged across all years	x	X					
$M_{L}16$	Median of annual minimum flows 2	Median of the lowest annual daily flows divided by median annual daily flows averaged across all years		x			x	x	
$M_{L}17$	Baseflow index 1	Seven-day min mum flow divided by mean annual daily flows averaged across all years						x	x
$M_{L}18$	Variability in Baseflow Index 1	Coefficient of variation in $M_{L}17$							x
$M_{L}21$	Variability across annual minimum flows	Coefficient of variation in annual minimum flows averaged across all years							x
$M_{L}22$	Specific mean annual minimum flows	Mean annual minimum flows divided by catchment area		x	x				
Magnitude of Flow Events: High Flow Conditions									
$M_{H}1$	Mean maximum monthly flows	Mean of the maximum monthly flows for all months			x				
$M_{H}4$	Mean maximum monthly flows	Mean of the maximum monthly flows for all months		x					
$M_{H}7$	Mean maximum monthly flows	Mean of the maximum monthly flows for all months		x					
$M_{H}8$	Mean maximum monthly flows	Mean of the maximum monthly flows for all months						x	x
$M_{H}9$	Mean maximum monthly flows	Mean of the maximum monthly flows for all months	x						

		Definition	Intermittent: Harsh Intermittent	Intermittent: Intermittent Flashy or Runoff	Perennial: Snowmelt	Perennial: Snow & Rain	Perennial: Superstable or Stable Groundwater	Perennial: Perennial Flashy or Runoff	All Streams
M_H10	Mean maximum monthly flows	Mean of the maximum monthly flows for all months					x		x
M_H14	Median of annual maximum flows	Median of the highest annual daily flow divided by the median annual daily flow averaged across all years	x	x				x	x
M_H16	High flow discharge	Mean of the 1st, 10th and 25th percentile from the flow duration curve divided by median daily flow across all years							x
M_H17	High flow discharge	Mean of the 1st, 10th and 25th percentile from the flow duration curve divided by median daily flow across all years				x	x		
M_H19	Skewness in annual maximum flows	See Hughes and James (1989)					x		
M_H20	Specific mean annual maximum flows	Mean annual maximum flows divided by catchment area			x	x			
M_H23	High flow volume	Mean of the high flow volume (calculated as the area between the hydrograph and the upper threshold during the high flow event) divided by median annual daily flow across all years. The upper threshold is defined as 1, 3, and 7 times median annual flow	x	x				x	
Frequency of Flow Events: Low Flow Conditions									
F_L1	Low flood pulse count	Number of annual occurrences during which the magnitude of flow remains below a lower threshold. Hydrologic pulses are defined as those periods within a year in which the flow drops below the 25th percentile (low pulse) of all daily values for the time period	x	x			x		x
F_L2	Variability in low Flood pulse count	Coefficient of variation in F_L1	x	x	x	x	x	x	x
F_L3	Frequency of low flow spells	Total number of low flow spells (threshold equal to 5% of mean daily flow) divided by the record length in years	x	x	x	x	x	x	x
Frequency of Flow Events: High Flow Conditions									
F_H2	Variability in high flood pulse count 1	Coefficient of variation in F_H1	x	x					x
F_H3	High flood pulse count 2	See F_H1, where the upper threshold is defined as 3 and 7 times median daily flow, and the value is represented as an average instead of a tabulated count		x		x	x		x
F_H4	High flood pulse count 2	See F_H1, where the upper threshold is defined as 3 and 7 times median daily flow, and the value is represented as an average instead of a tabulated count						x	

		Definition	Intermittent		Perennial				
			Harsh Intermittent	Intermittent Flashy or Runoff	Snowmelt	Snow & Rain	Superstable or Stable Groundwater	Perennial Flashy or Runoff	All Streams
F_H5	Flood frequency 1	Mean number of high flow events per year using an upper threshold of 1, 3, and 7 times median flow over all years	x			x			
F_H6	Flood frequency 1	Mean number of high flow events per year using an upper threshold of 1, 3, and 7 times median flow over all years					x	x	x
F_H7	Flood frequency 1	Mean number of high flow events per year using an upper threshold of 1, 3, and 7 times median flow over all years	x	x				x	x
F_H8	Flood frequency 2	See F_H5–7, where the 25th and 75th percentile are used as the upper threshold			x				
F_H10	Flood frequency 3	See F_H5–7, where the median of the annual minima is used as the upper threshold		x					
F_H11	Flood frequency 4	Mean number of discrete flood events per year			x		x		
Duration of Flow Events: Low Flow Conditions									
D_L1	Annual minima of 1-/3-/7-/30-/90-day means of daily discharge	Magnitude of minimum annual flow of various duration, ranging from daily to seasonal	x	x					
D_L2	Annual minima of 1-/3-/7-/30-/90-day means of daily discharge	Magnitude of minimum annual flow of various duration, ranging from daily to seasonal	x						
D_L5	Annual minima of 1-/3-/7-/30-/90-day means of daily discharge	Magnitude of minimum annual flow of various duration, ranging from daily to seasonal			x				
D_L6	Variability in annual minima of one-/3-/7-/30-/90-day means of daily discharge	Coefficient of variation in D_L1-5				x		x	
D_L9	Variability in annual minima of one-/3-/7-/30-/90-day means of daily discharge	Coefficient of variation in D_L1-5					x		
D_L10	Variability in annual minima of one-/3-/7-/30-/90-day means of daily discharge	Coefficient of variation in D_L1-5						x	
D_L11	Means of 1-/7-/30-/90-day minima of daily discharge	Mean annual 1-day/7-day/30-day minimum, respectively, divided by median flow					x		
D_L13	Means of 1-/7-/30-/90-day minima of daily discharge	Mean annual 1-day/7-day/30-day minimum, respectively, divided by median flow	x	x		x			x
D_L16	Low flow pulse duration	Mean duration of F_L1		x	x		x		x
D_L17	Variability in low flow pulse duration	Coefficient of variation in D_L16						x	x

		Definition	Intermittent		Perennial				
			Harsh Intermittent	Intermittent Flashy or Runoff	Snowmelt	Snow & Rain	Superstable or Stable Groundwater	Perennial Flashy or Runoff	All Streams
D_L18	Number of zero-flow days	Mean annual number of days having zero daily flow		x					x
Duration of Flow Events: High Flow Conditions									
D_H5	Annual maxima of 1-/3-/7-/30-/90-day means of daily discharge	Magnitude of maximum annual flow of various duration, ranging from daily to seasonal	x						
D_H10	Variability in annual maxima of 1-/3-/7-/30-day means of daily discharge	Coefficient of variation in D_H1-5	x						
D_H11	Means of 1-/7-/30-day maxima of daily discharge	Mean annual 1-day/7-day/30-day maximum, respectively, divided by median flow					x		
D_H12	Means of 1-/7-/30-day maxima of daily discharge	Mean annual 1-day/7-day/30-day maximum, respectively, divided by median flow		x		x			
D_H13	Means of 1-/7-/30-day maxima of daily discharge	Mean annual 1-day/7-day/30-day maximum, respectively, divided by median flow		x				x	x
D_H15	High flow pulse duration	Mean duration of F_H1		x			x		x
D_H16	Variability in high flow pulse duration	Coefficient of variation in D_H15			x			x	x
D_H19	High flow duration 1	See DH15, where the upper threshold is defined as 1, 3, and 7 times median flows, and the value is represented as an average instead of a tabulated count			x				
D_H20	High flow duration 2	See DH17–19, where the upper threshold is defined as the 25th and 75th percentile of median flows					x		x
D_H22	Flood interval	Mean annual median interval in days between floods over all years	x						
D_H23	Flood duration 2	Mean annual number of days that flows remain above the flood threshold averaged across all years		x					
D_H24	Flood free days	Mean annual maximum number of 365 days over all water years during which no floods occurred over all years				x		x	
Timing of Flow Events									
T_A1	Constancy	See Colwell (1974)		x	x	x	x	x	x
T_A2	Predictability of flow	Composed of two independent, additive components: constancy (a measure of temporal invariance) and contingency (a measure of periodicity)		x					
T_A3	Seasonal predictability of flooding	Maximum proportion of all floods over the period of record that fall in any one of six 60-day 'seasonal' windows			x			x	
T_L1	Julian date of annual minimum	The mean Julian date of the 1-day annual minimum flow over all years		x		x			

		Definition	Intermittent		Perennial				
			Harsh Intermittent	Intermittent Flashy or Runoff	Snowmelt	Snow & Rain	Superstable or Stable Groundwater	Perennial Flashy or Runoff	All Streams
T_L2	Variability in Julian date of annual minimum	Coefficient of variation in T_L1	x				x		x
T_H1	Julian date of annual maximum	The mean Julian date of the 1-day annual maximum flow over all years	x				x		
T_H2	Variability in Julian date of annual maximum	Coefficient of variation in T_H1	x						
T_H3	Seasonal predictability of non-flooding	Maximum proportion of the year (number of days/365) during which no floods have ever occurred over the period of record		x				x	x
Rate of Change in Flow Events									
R_A1	Rise rate	Mean rate of positive changes in flow from one day to the next	x		x				
R_A4	Variability in fall rate	Coefficient of variation in RA_3	x						
R_A5	No day rises	Ratio of days where the flow is higher than the previous day	x	x			x		x
R_A6	Change of flow	Median of difference between natural logarithm of flows between two consecutive days with increasing/decreasing flow		x				x	x
R_A7	Change of flow	Median of difference between natural logarithm of flows between two consecutive days with increasing/decreasing flow		x				x	
R_A8	Reversals	Number of negative and positive changes in water conditions from one day to the next			x	x	x		x
R_A9	Variability in reversals	Coefficient of variation in RA_8		x		x	x	x	x
R_A1	Rise Rate	Mean rate of positive changes in flow from one day to the next	x		x				
R_A4	Variability in fall rate	Coefficient of variation in RA_3	x						
R_A5	No day rises	Ratio of days where the flow is higher than the previous day	x	x			x		x

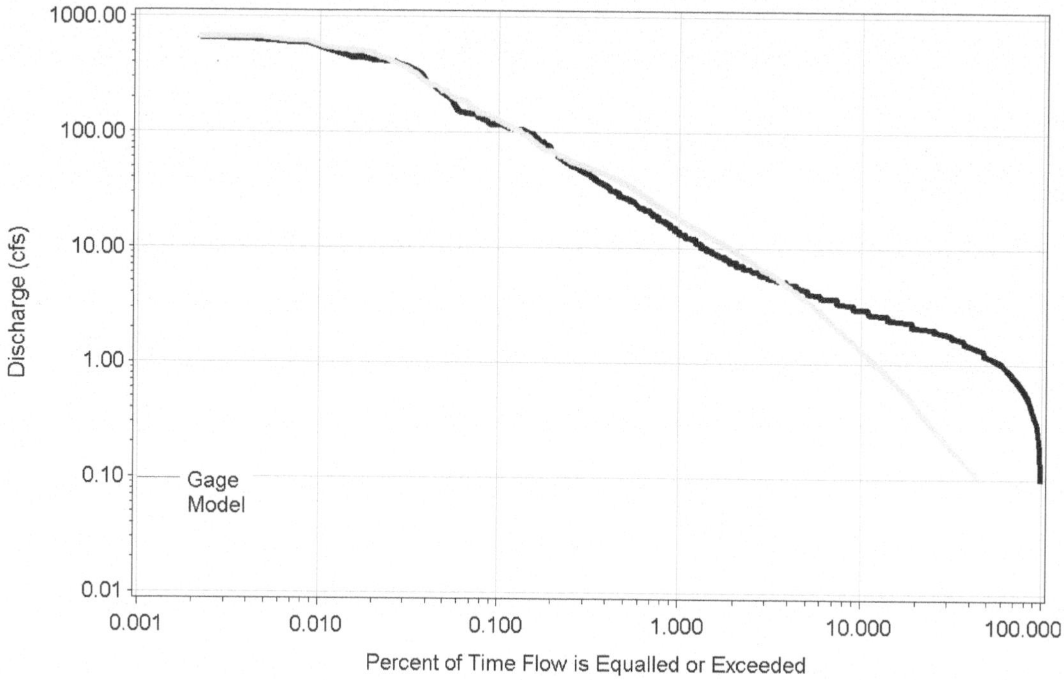

Figure A-1. Flow Duration Curve For Calibration Period: Camp Creek.

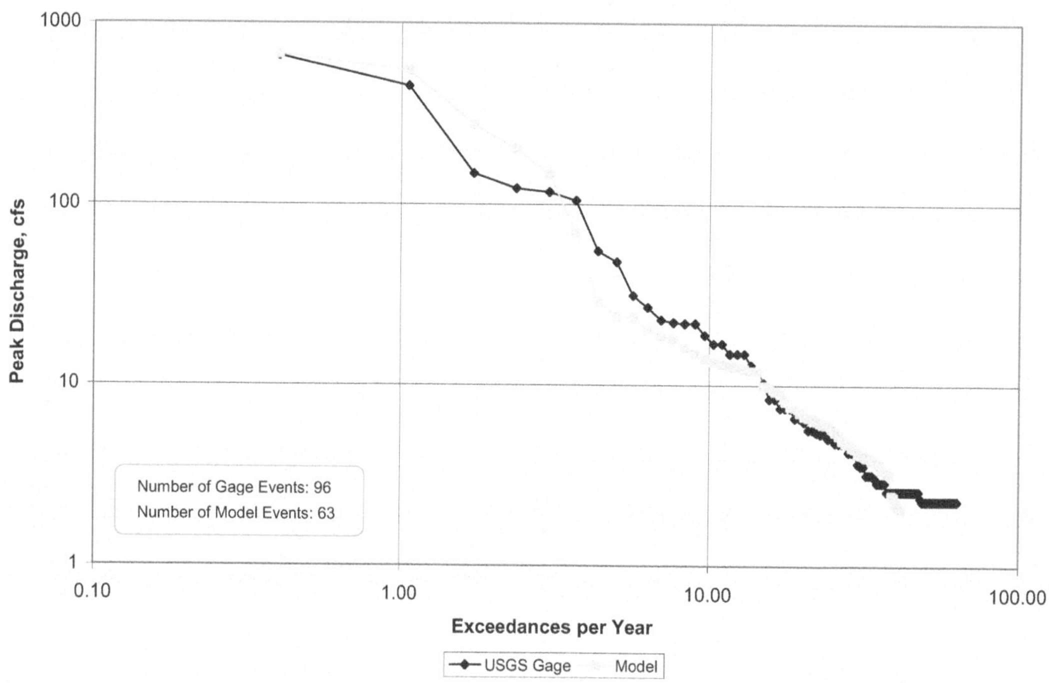

Figure A-2. Peak Flow Frequency Exceedance Curve For Calibration Period: Camp Creek.

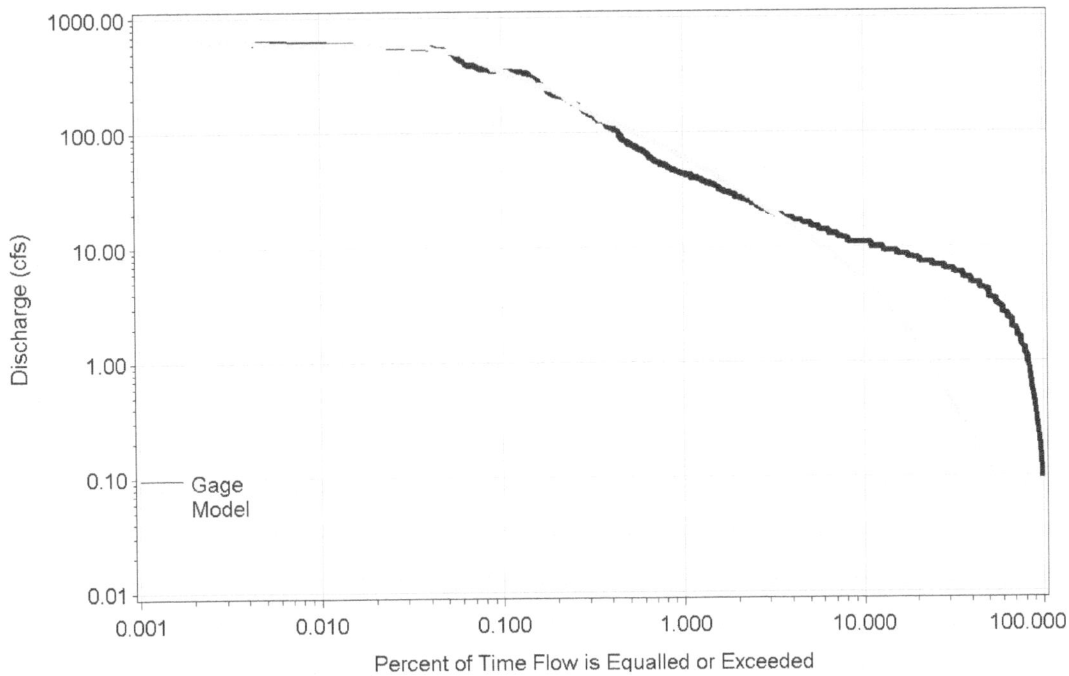

Figure A-3. Flow Duration Curve For Calibration Period: Wilson Creek.

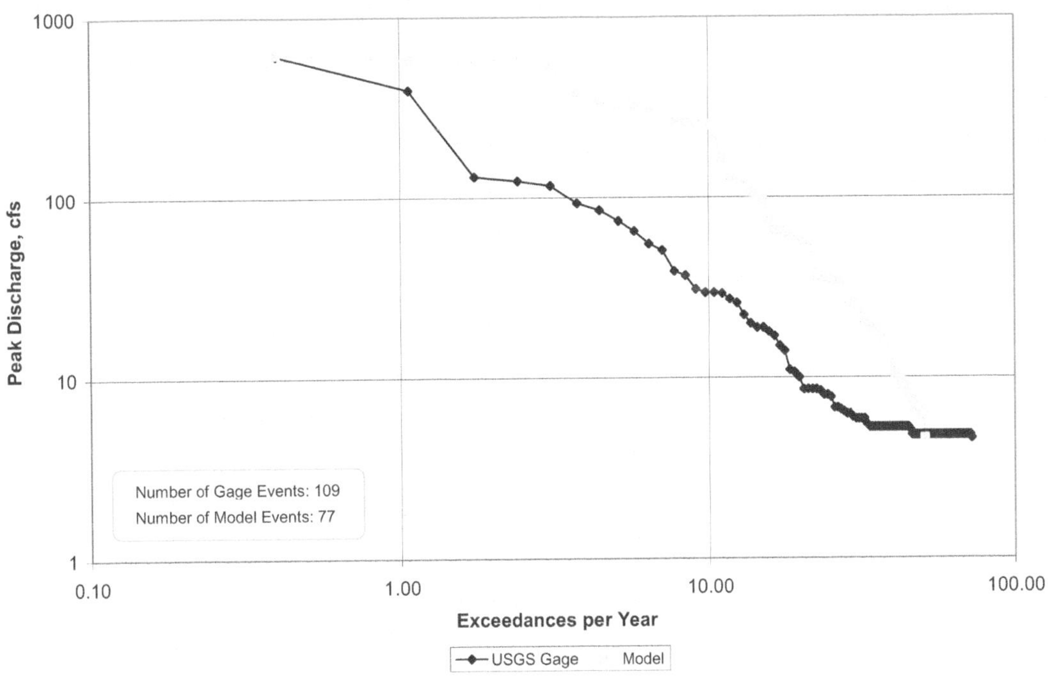

Figure A-4. Peak Flow Frequency Exceedance Curve For Calibration Period: Wilson Creek.

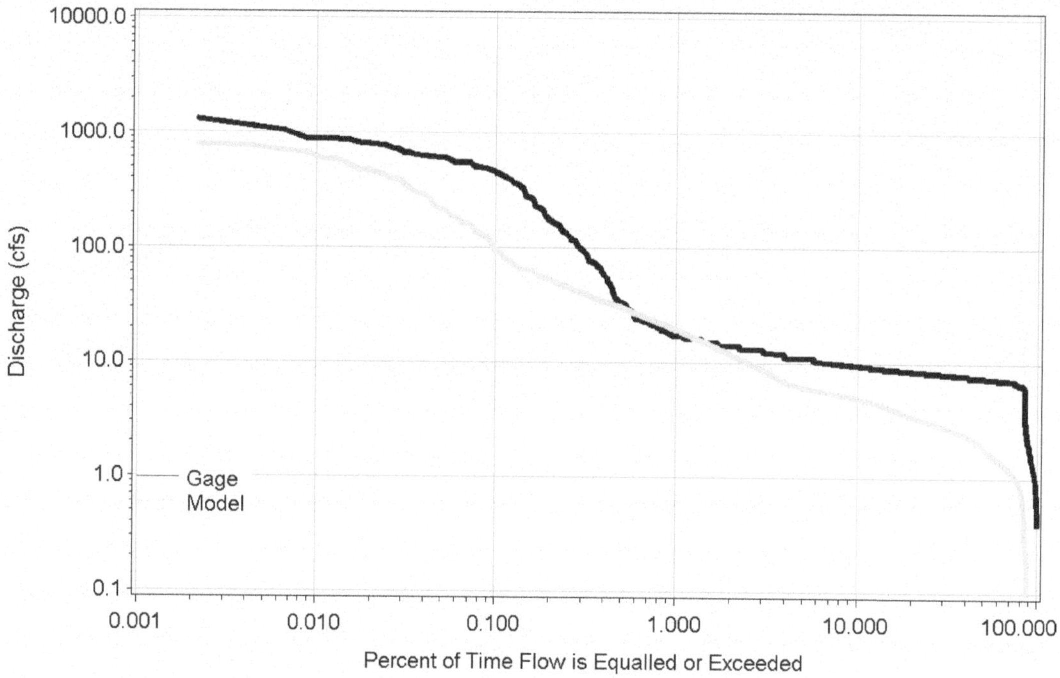

Figure A-5. Flow Duration Curve For Calibration Period: Dutchmans Creek.

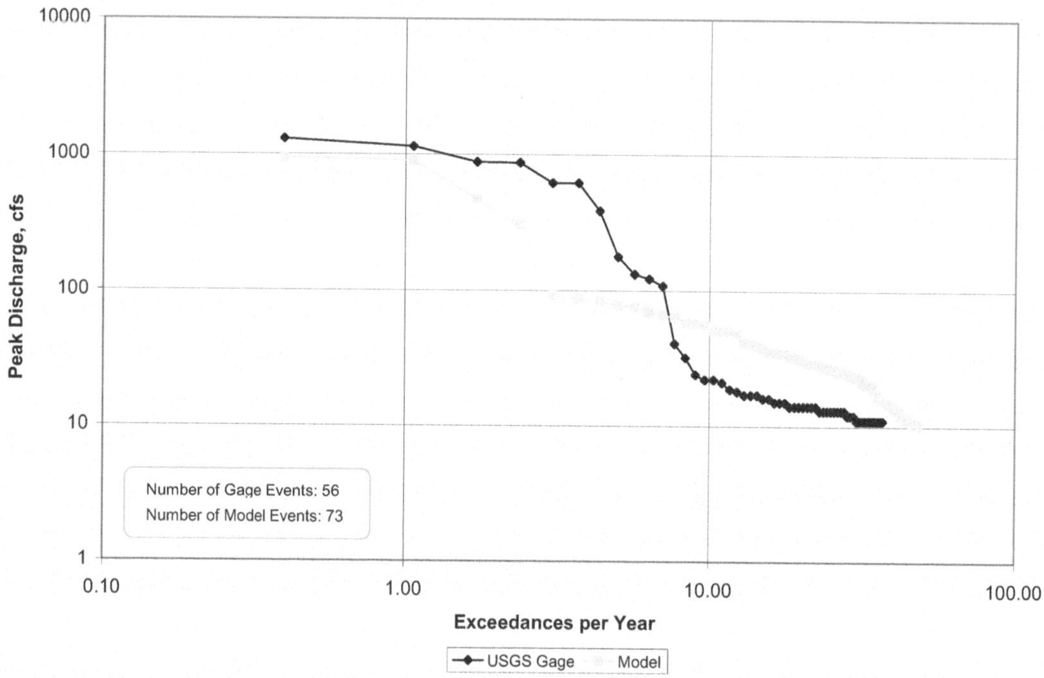

Figure A-6. Peak Flow Frequency Exceedance Curve For Calibration Period: Dutchmans Creek.

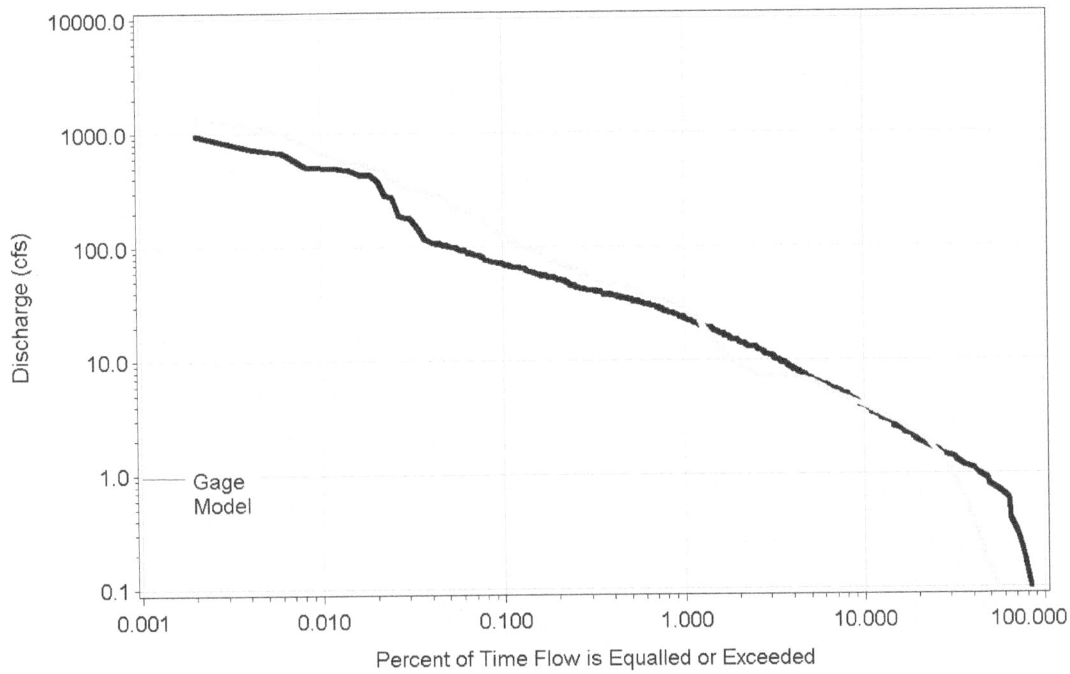

Figure A-7. Flow Duration Curve For Calibration Period: Black Creek.

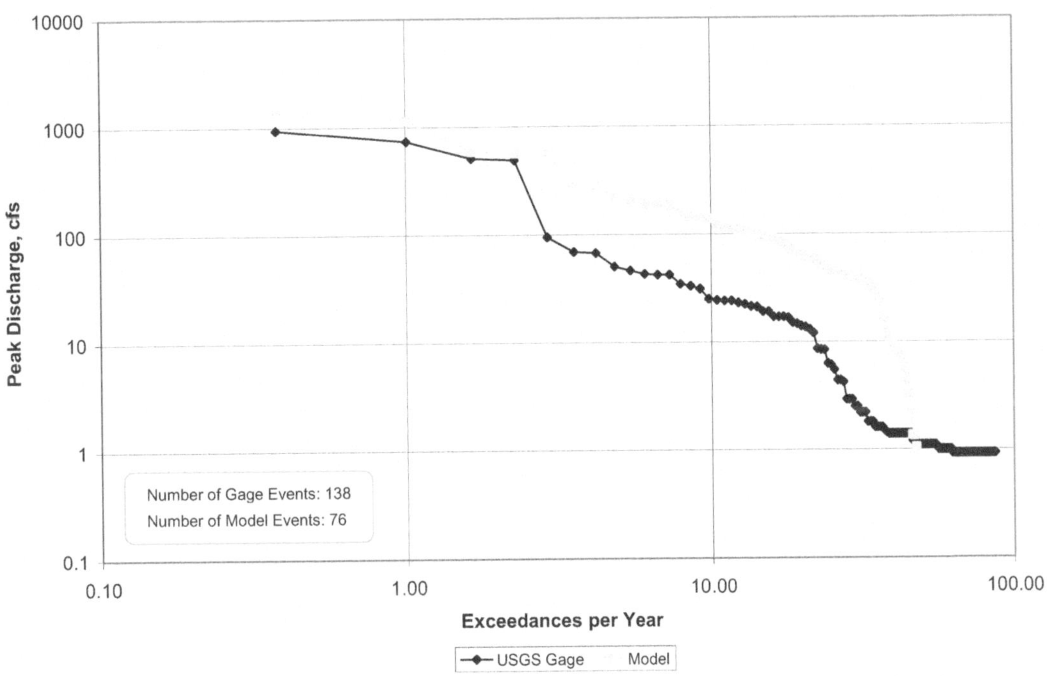

Figure A-8. Peak Flow Frequency Exceedance Curve For Calibration Period: Black Creek.

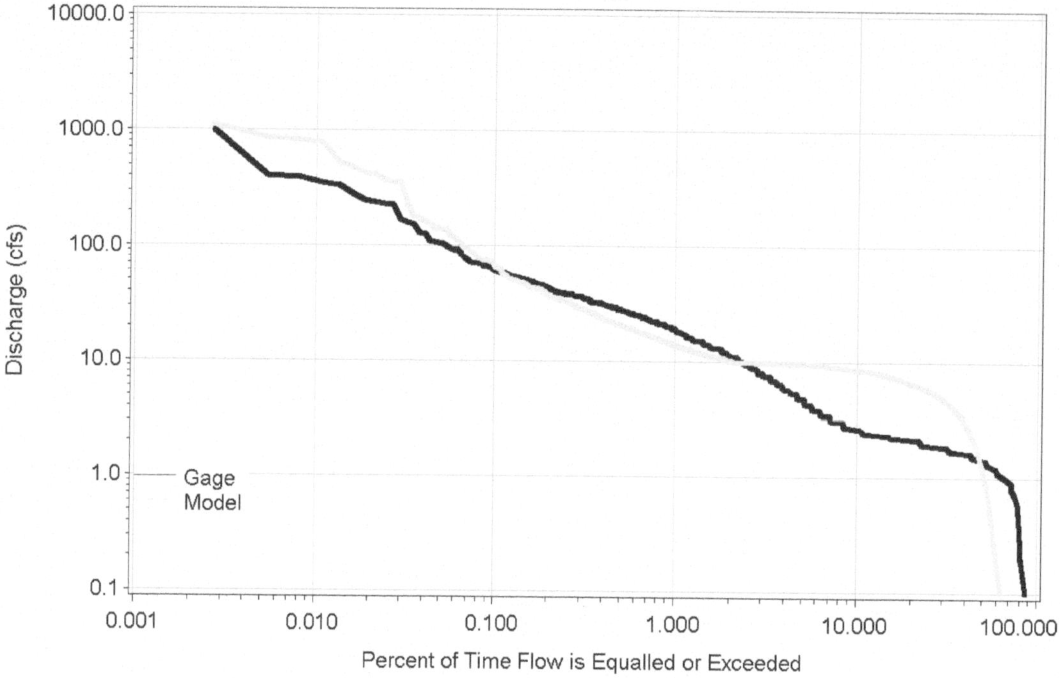

Figure A-9. Flow Duration Curve For Calibration Period: Beaverdam Creek.

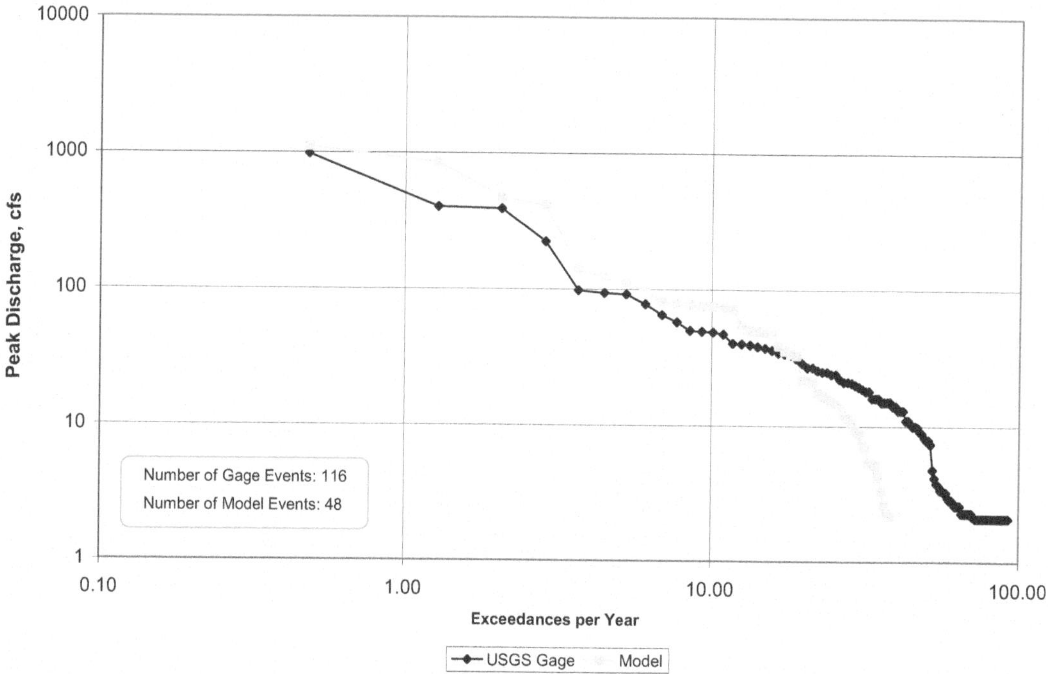

Figure A-10. Peak Flow Frequency Exceedance Curve For Calibration Period: Beaverdam Creek.

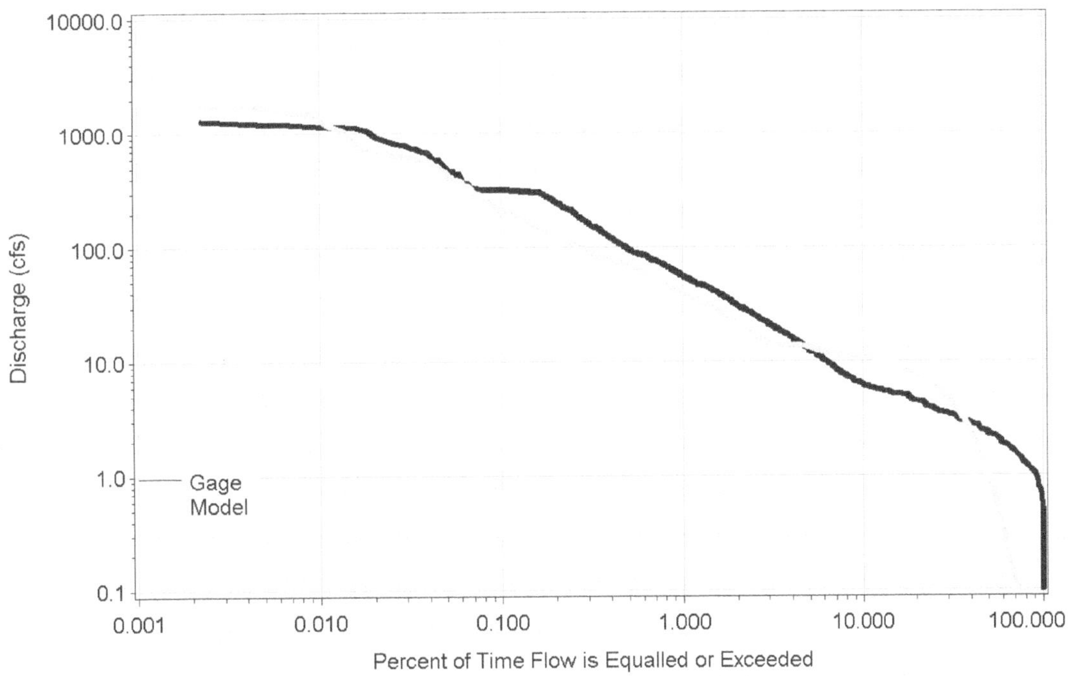

Figure A-11. Flow Duration Curve For Calibration Period: Pigeon House Creek.

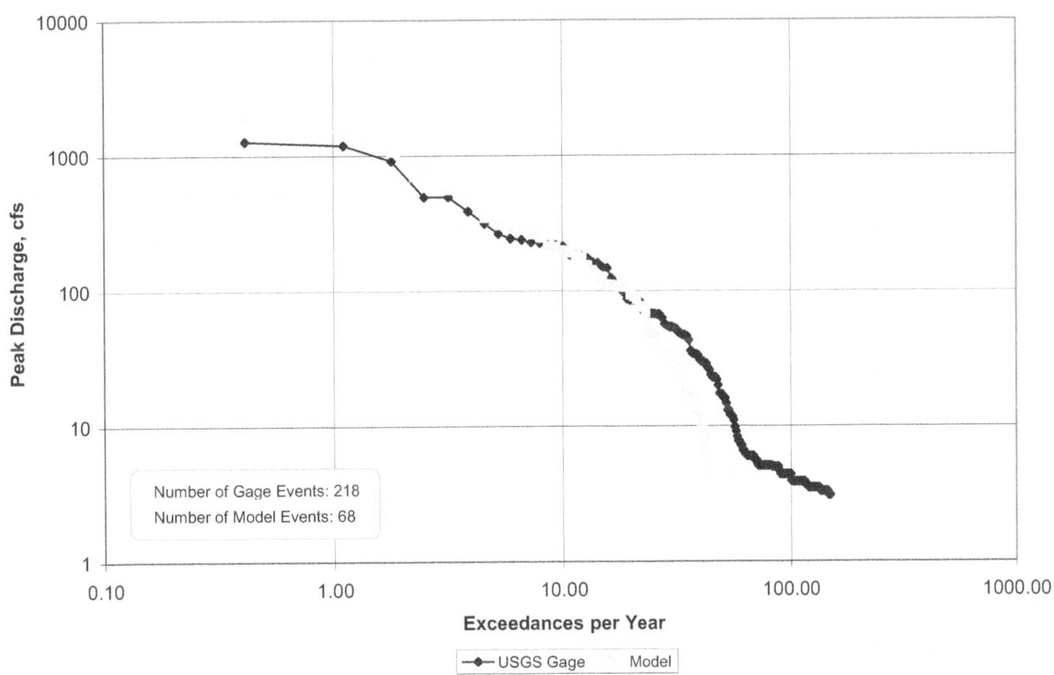

Figure A-12. Peak Flow Frequency Exceedance Curve For Calibration Period: Pigeon House Creek.

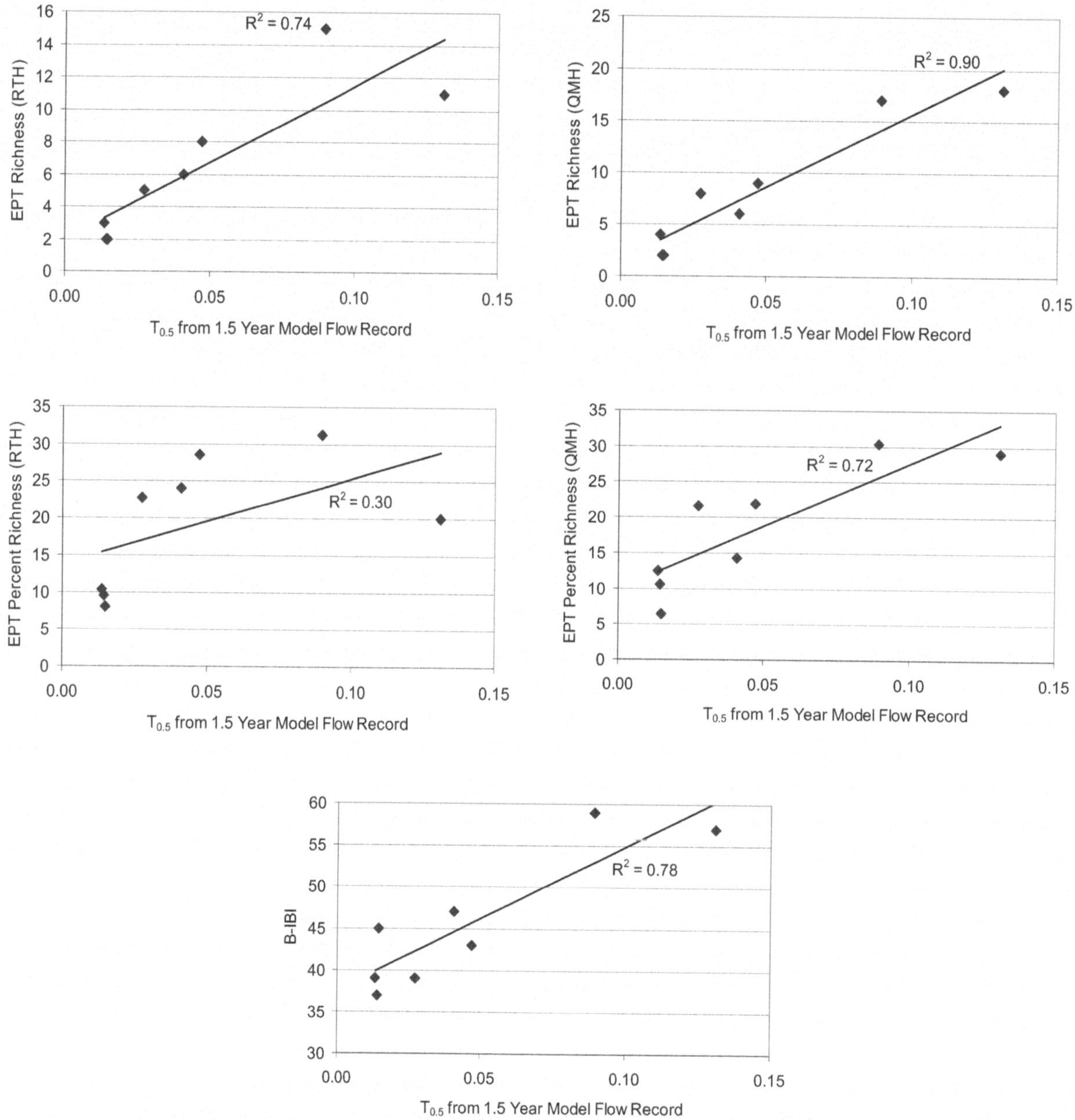

Figure A-13. Relationship of $T_{0.5}$ Calculated From the Calibration Period Flow Record to Benthic Macroinvertebrate Sampling Data (a) EPT Richness (RTH), (b) EPT Richness (QMH), (c) EPT Percent Richness (RTH), (d) EPT Percent Richness (QMH), (e) Benthic Index of Biotic Integrity (B-IBI).

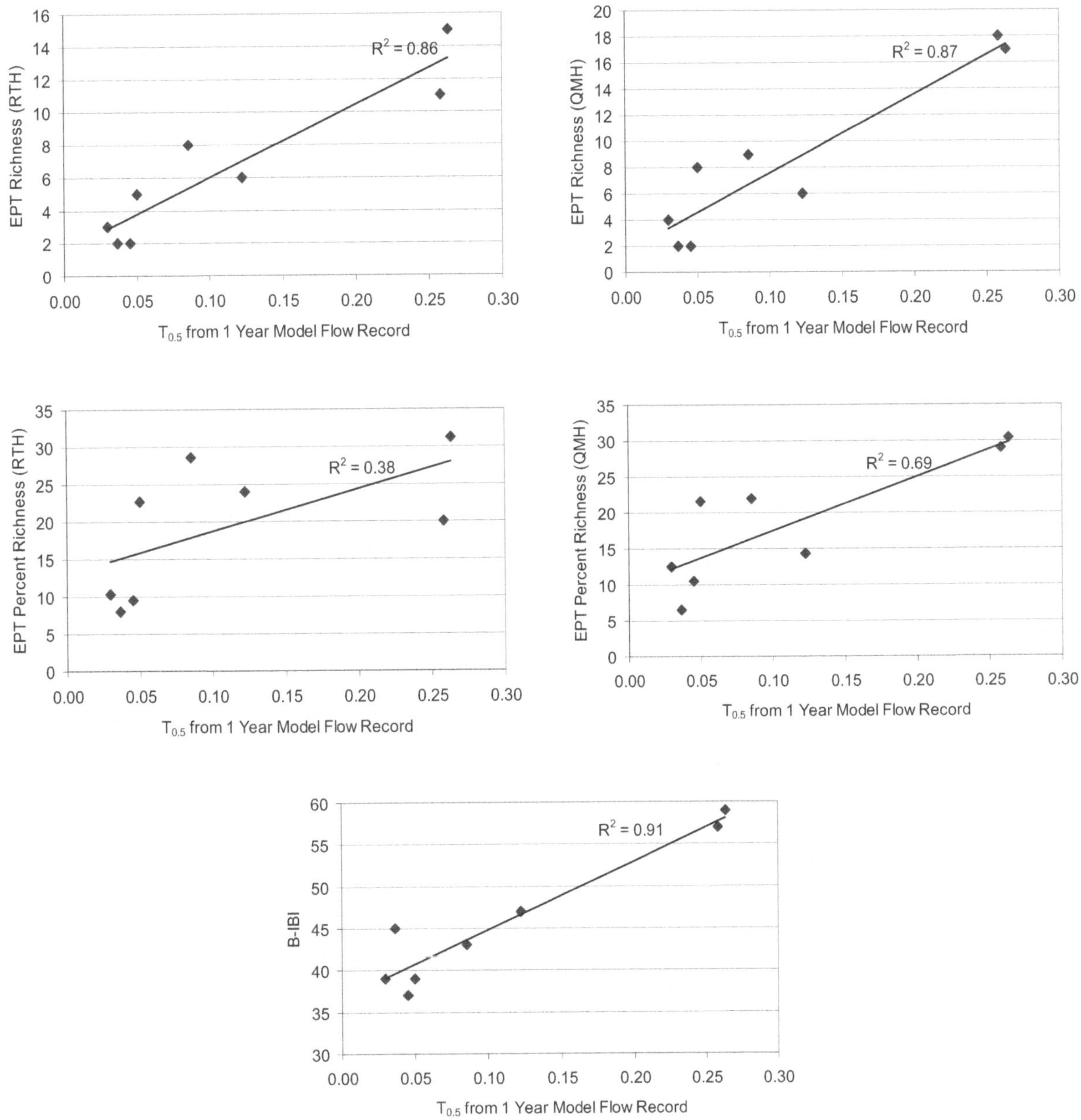

Figure A-14. Relationship of $T_{0.5}$ Calculated From the 1-Year Flow Record to Benthic Macroinvertebrate Sampling Data (a) EPT Richness (RTH), (b) EPT Richness (QMH), (c) EPT Percent Richness (RTH), (d) EPT Percent Richness (QMH), (e) Benthic Index of Biotic Integrity (B-IBI).

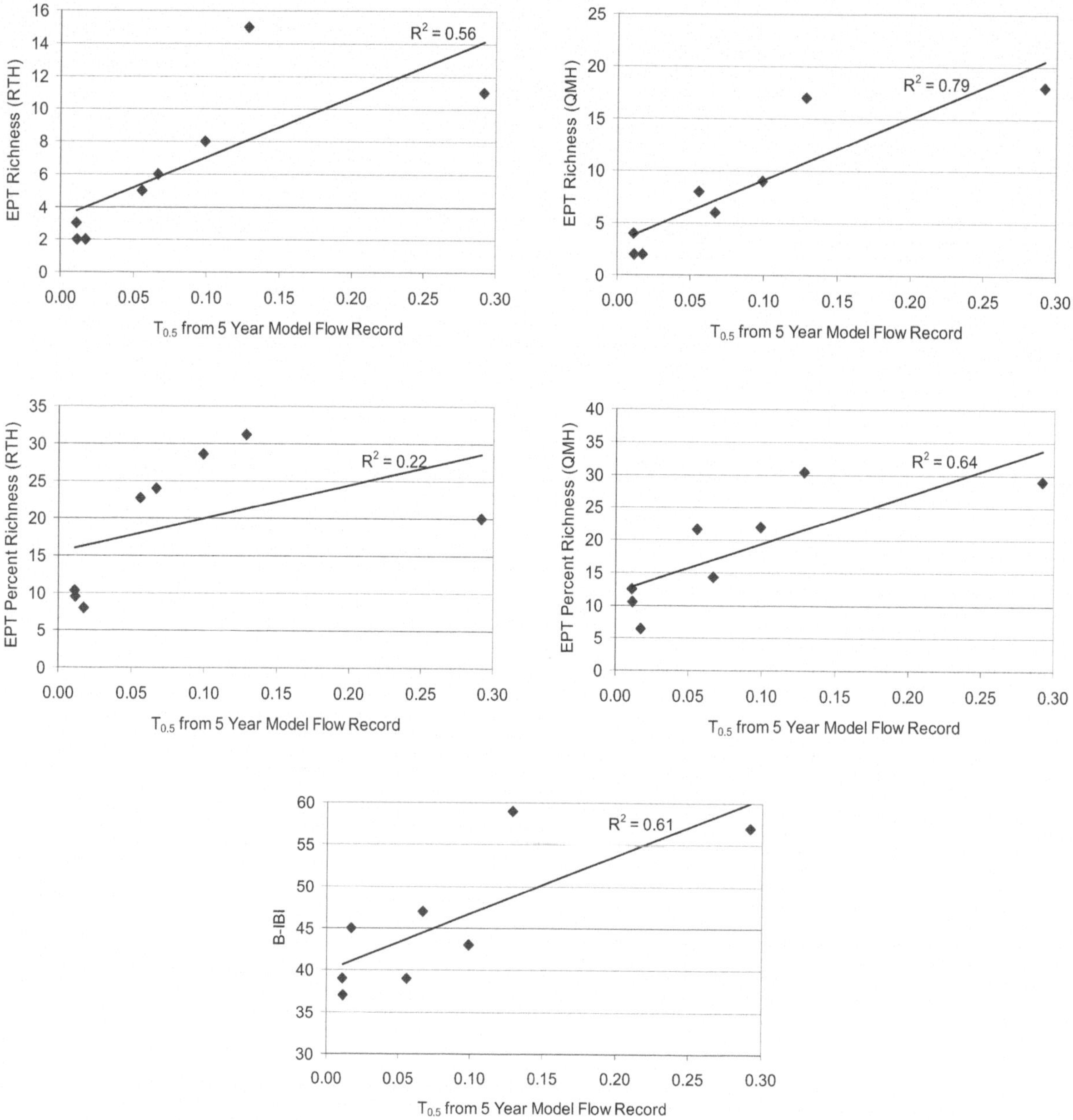

Figure A-15. Relationship of $T_{0.5}$ Calculated From the 5-Year Flow Record to Benthic Macroinvertebrate Sampling Data (a) EPT Richness (RTH), (b) EPT Richness (QMH), (c) EPT Percent Richness (RTH), (d) EPT Percent Richness (QMH), (e) Benthic Index of Biotic Integrity (B-IBI).

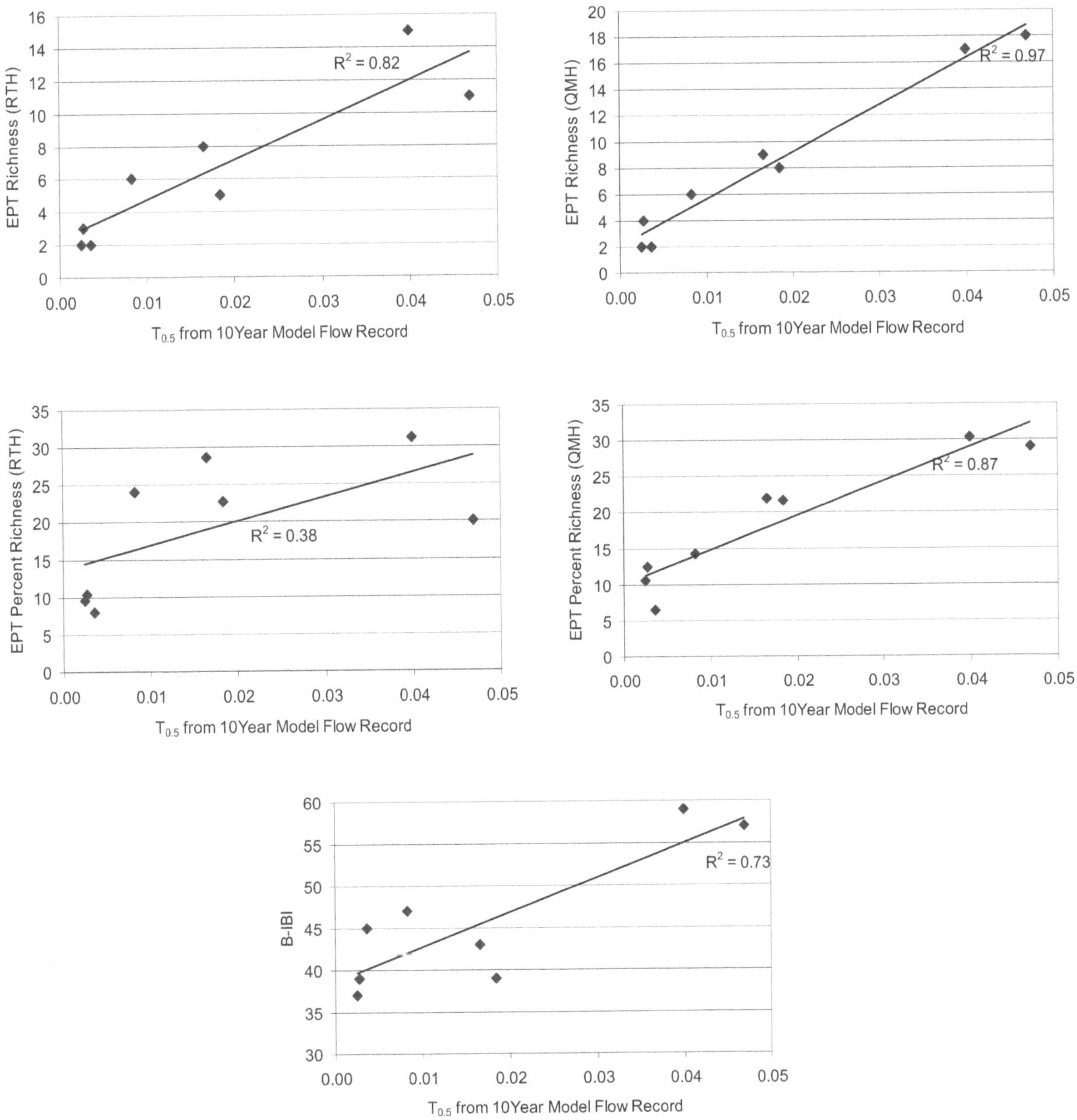

Figure A-16. Relationship of $T_{0.5}$ Calculated From the 10-Year Flow Record to Benthic Macroinvertebrate Sampling Data (a) EPT Richness (RTH), (b) EPT Richness (QMH), (c) EPT Percent Richness (RTH), (d) EPT Percent Richness (QMH), (e) Benthic Index of Biotic Integrity (B-IBI).

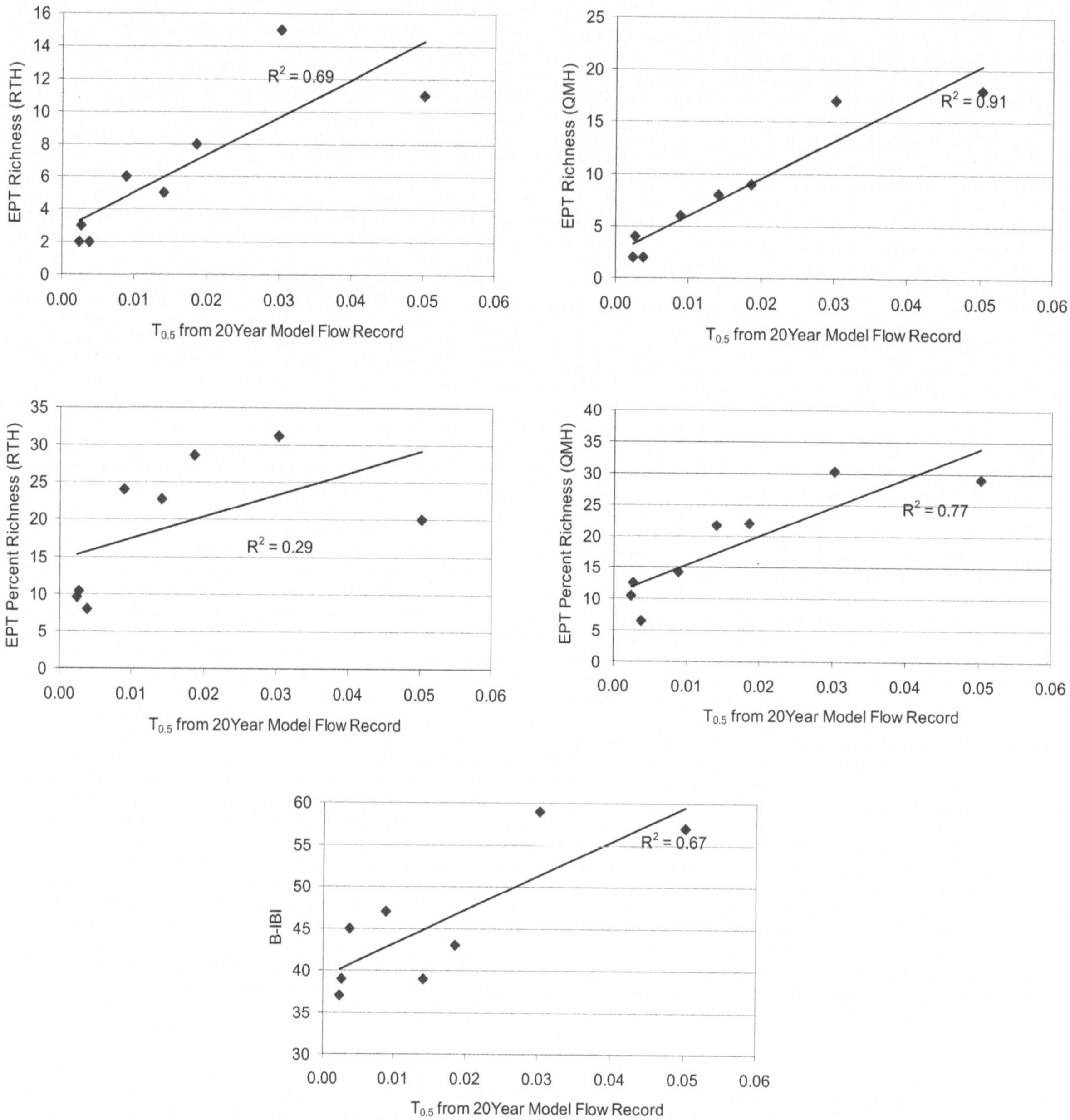

Figure A-17. Relationship of $T_{0.5}$ Calculated From the 20-Year Flow Record to Benthic Macroinvertebrate Sampling Data (a) EPT Richness (RTH), (b) EPT Richness (QMH), (c) EPT Percent Richness (RTH), (d) EPT Percent Richness (QMH), (e) Benthic Index of Biotic Integrity (B-IBI).

References

Akan, A.O. and Houghtalen, R.J. 2003. *Urban Hydrology, Hydraulics, and Stormwater Quality: Engineering Applications and Computer Modeling.* John Wiley & Sons, New York.

Alberti, M., Booth, D.B., Hill, M., and Marzluff, J. 2003. *The Impacts of Urban Patterns on Ecosystem Dynamics.* Final report. National Science Foundation, Arlington, VA.

Allan, C. and Curtis, A. 2005. Nipped in the Bud: Why Regional Scale Adaptive Management Is Not Blooming. *Environmental Management* 36(3):414–425.

Arcement, G.J. and Schneider, V.R. 1989. Guide for Selecting Manning's Roughness Coefficients for Natural Channels and Flood Plains. U.S. Geological Survey Water Supply Paper 2339.

Armentrout, N.B., compiler. 1998. *Glossary of Aquatic Habitat Inventory Terminology.* American Fisheries Society, Bethesda, MD.

Bain, M.B. and Stevenson, N.J., eds. 1999. *Aquatic Habitat Assessment: Common Methods.* American Fisheries Society, Bethesda, MD.

Baker, D.B., Richards, R.P., Loftus, T.T., and Kramer, J.W. 2004. A New Flashiness Index: Characteristics and Applications to Midwestern Rivers and Streams. *American Water Resources Association* 40:503–522.

Barbour M.T. and Yoder, C.O. 2007. *Technical Guidelines: Critical Technical Elements for an Effective Bioassessment Program.*

Barbour, M.T., Gerritsen, J., Snyder, B.D., and Stribling, J.B. 1999. *Rapid Bioassessment Protocols for Use in Streams and Wadeable Rivers: Periphyton, Benthic Macroinvertebrates and Fish.* 2nd edition. EPA 841-B-99-002. U.S. Environmental Protection Agency, Office of Water, Washington, D.C.

Barbour, M.T., Paul, M.J., Bressler, D.W., Purcell, A.H., Resh, V.H., and Rankin, E. In press. *Research Digest: Bioassessment: A Tool for Managing Aquatic Life Uses for Urban Streams.* Water Environmental Research Foundation Project No. 01-WSM-3.

Bartleya, R. and Rutherfurd, I. 2005. Measuring the Reach-Scale Geomorphic Diversity of Streams: Application to a Stream Disturbed by a Sediment Slug. *River Research and Applications* 21:39–59.

Berry, W., Rubinstein, N. Melzian, B., and Hill, B. 2003. *The Biological Effects of Suspended and Bedded Sediment (SABS) in Aquatic Systems: A Review.* U.S. EPA Internal Report, August 20, 2003. Office of Research and Development.

Bevenger, G.S. and King, R.M. 1995. *A Pebble Count Procedure for Assessing Watershed Cumulative Effects.* Research Paper RM-RP-319. Fort Collins, CO: U.S. Department of Agriculture, Forest Service, Rocky Mountain Forest and Range Experiment Station.

Bicknell, B.R., Imhoff, J.C., Kittle, J.L., Jr., Donigian, A.S., Jr., and Johanson, R.C. *Hydrologic Simulation Program – Fortran: User's Manual for Release 11*, U.S. EPA, Office of Research and Development, Athens, GA, 1997. Available from: www.epa.gov/ceampubl/softwdos.htm.

Bledsoe, B.P. 2002. Stream Erosion Potential and Stormwater Management Strategies. *Journal of Water Resources Planning and Management*, 128:451–455.

Bledsoe, B.P. and Watson, C.C. 2001. Effects of Urbanization on Channel Instability. *Journal of the American Water Resources Association* 37(2):225–270.

Bledsoe, B.P., Brown, M.C., and Raff, D.A. In press. GeoTools: A Toolkit for Fluvial System Analysis. *Journal of the American Water Resources Association.*

Booker, D.J. and Dunbar, M.J. 2004. Application of Physical Habitat Simulation (PHABSIM) Modeling to Modified Urban River Channels. *River Research and Applications* 20(2):167–183.

Booth, D.B., Hartley, D., and Jackson, R. 2002. Forest Cover, Impervious-Surface Area, and the Mitigation of Stormwater Impacts. *Journal of the American Water Resources Association* 38(3): 835–845.

Booth, D.B. and Henshaw, P.C. 2001. Rates of Channel Erosion in Small Urban Streams. In: Wigmosta, M.S., and S.J. Burgers (Editors), *Land Use and Watershed, Human Influence on Hydrology and Geomorphology in Urban and Forest Areas.* Washington, D.C.: AGU, 17–38.

Booth, D.B. and Jackson, C.R. 1997. Urbanization of Aquatic Systems: Degradation Thresholds, Stormwater Detection, and the Limits of Mitigation. *Journal of the American Water Resources Association* 33(5):1077–1090.

Booth, D.B., Karr, J.R., Schauman, S., Konrad, C.P., Morley, S.A., Larson, M.G., and Burger, S.J. 2004. Reviving Urban Streams: Land Use, Hydrology, Biology, and Human Behavior. *Journal of the American Water Resources Association* 40(5):1351–1364.

Booth, D.B., Karr. J.R., Schauman, S., Konrad, C.P., Morley, S.A., Larson, M.G., and Burges, S.J. 2003. Management Strategies For Urban Stream Rehabilitation. *Proceedings of the National Conference on Urban Stormwater: Enhancing Programs at the Local Level,* Chicago. February 17–20, 2003. EPA/625/R-03/003.

Borsuk, M.E., Stow, C.A. and Reckhow, K.H. 2004. Integrative Environmental Prediction Using Bayesian Networks: A Synthesis of Models Describing Estuarine Eutrophication. *Ecological Modeling.* 173(2–3):219–239.

Bovee, K.D., Lamb, B.L., Bartholow, J.M., Stalnaker, C.B., Taylor, J., and Henriksen, J. 1998. Stream Habitat Analysis Using the Instream Flow Incremental Methodology. *Information and Technology Report USGS/BRD-1998-0004.* U.S. Geological Survey, Biological Research Division, Washington, D.C.

Brabec, E., Schulte, S., and Richards, P.L. 2002. Impervious Surfaces and Water Quality: A Review of Current Literature and Its Implications for Watershed Planning. *Journal of Planning Literature* 16(4):499–514.

Bunn, S.E. and Arthington, A.H. 2002. Basic Principles and Ecological Consequences of Altered Flow Regimes for Aquatic Biodiversity. *Environmental Management* 30(4):492–507.

Bunte, K. and Abt, S.R. 2001. *Sampling Surface and Subsurface Particle-Size Distributions in Wadable Gravel- and Cobble-Bed Streams for Analysis in Sediment Transport, Hydraulics, and Streambed Monitoring.* United States Department of Agriculture General Technical Report RMRS-GTR-74. United States Department of Agriculture, Forest Service.

Burton, Jr., G.A. 2002. Sediment Quality Criteria in Use Around the World. *Limnology* 3(2):65–76.

Cassin, J., Fuerstenberg, R., Tear, L., Whiting, K., St. John, D., Murray, B., and Burkey, J. 2005. Development of Hydrological and Biological Indicators of Flow Alteration in Puget Sound Lowland Streams. Seattle, WA: King County Water and Land Resources Division.

Center for Watershed Protection (CWP). 2002. Is Impervious Cover Still Important? Section 1: A Review of Recent Urban Stream Research. www.cwp.org/Impacts_Part1.PDF. Ellicott City, MD: Author.

Center for Watershed Protection (CWP). 2003. Is Impervious Cover Still Important? Section 2: Impacts of Urbanization on Downstream Receiving Waters. www.cwp.org/Impacts_Part2.PDF. Ellicott City, MD: Author.

Chessman, B.C., Fryirsb, K.A., and Brierley, G.J. 2006. Linking Geomorphic Character, Behaviour and Condition to Fluvial Biodiversity: Implications for River Management. *Aquatic Conservation: Marine and Freshwater Ecosystems* 16:267–288.

Chow, V.T. 1959. Open Channel Hydraulics. McGraw-Hill, New York.

Clausen, B. and Biggs, B.J.F. 2000. Flow Variables for Ecological Studies in Temperate Streams: Groupings Based on Covariance. *Journal of Hydrology* 237(3–4):184–197.

Clausen, B. and Biggs, B.J.F. 1997. Relationships between Benthic Biota and Hydrological Indices in New Zealand Streams. *Freshwater Biology* 38(2):327–342.

Coffman, L.S. 2003. *Discussion of an Ecosystem Functional Basis for Protecting Receiving Waters. Protection and Restoration of Streams.* 139, 10. 2003/06/23/. American Society of Civil Engineers, Philadelphia.

Coles, J.F., Cuffney, T.F., and McMahon, G. 2004. The Effects of Urbanization on the Biological, Physical, and Chemical Characteristics of Coastal New England Streams. U.S. Geological Survey Professional Paper No. 1695.

Couch, C. and Hamilton, P. (2002). Effects of Urbanization on Stream Ecosystems. U.S. Geological Survey Fact Sheet FS-042-02.

Cunnane, C. 1978. Unbiased Plotting Positions: A Review. *Journal of Hydrology*. 37:205–222.

Davenport, A.J., Gurnella, A.M., and Armitage, P.D. 2004. Habitat Survey and Classification of Urban Rivers. *River Research and Applications* 20:687–704.

Davies, S.P. and Jackson, S.K. 2006. The Biological Condition Gradient: A Descriptive Model for Interpreting Change in Aquatic Ecosystems. *Ecological Applications,* 16(4):1251–1266.

Dytham, C. 1999. *Choosing and Using Statistics: A Biologist's Guide.* Blackwell Science, Malden, MA.

Edgerly, J.L. 2006. Quantifying Urban-Induced Flow Regime Alteration Using Mathematical Models and Hydrologic Metrics. Thesis, Department of Civil Engineering, Colorado State University, Fort Collins, CO.

Edgerly, J.L., Roesner, L.A., Rohrer, C.A., and Gironas, J.A. 2006. Quantifying Urban-induced Flow Regime Alteration and Evaluating Mitigation Alternatives Using Mathematical Models and Hydrologic Metrics. World Environmental and Water Resources Congress: Examining the Confluence of Environmental and Water Concerns. Omaha, NE, May 21–25.

Eisele, M., Steinbrich, A., Hildebrand, A., and Leibundgut, C. 2003. The Significance of Hydrological Criteria for the Assessment of the Ecological Quality in River Basins. *Physics and Chemistry of the Earth* 28(12–13):529–536.

Erskine, W.D., Terrazzolo, N., and Warner, R.F. 1999. River Rehabilitation from the Hydrologic and Geomorphic Impacts of a Large Hydro-Electric Power Project: Snowy River, Australia. *Regulated Rivers Research & Management* 15:3–24.

Finkenbine, J.K., Atwater, J.W., and Mavinic, D.S. 2000. Stream Health after Urbanization. *Journal of the American Water Resources Association* 36(5):1149–1160.

Fitzpatrick, F.A. and Peppler, M.C. 2007. Changes in Aquatic Habitat and Geomorphic Response to Urbanization, with Implications for Assessing Habitat Degradation. EWRI 2007. *Restoring our Natural Habitat. Proceedings of the 2007 World Environmental and Water Resources Congress.* May 15–19, 2007. Tampa, FL.

Fitzpatrick, F.A., Waite, I.R., D'Arconte, P.J., Meador, M.R., Maupin, M.A., and Gurtz, M.E. 1998. *Revised Methods for Characterizing Stream Habitat in the National Water Quality Assessment Program, Water-Resources Investigations Report 98-4052.* U.S. Geological Survey. Raleigh, NC.

Fitzpatrick, F.A., Diebel, M.W., Harris, M.A., Arnold, T.L., Lutz, M.A. and Richards, K.D. 2005. Effects of Urbanization on the Geomorphology, Habitat, Hydrology, and Fish Index of Biotic Integrity of Streams in the Chicago Area, Illinois, and Wisconsin. American Fisheries Society Symposium. 47:87–115.

Flotemersch, J.E., Cormier, S.M., and Autrey, B.C. 2001. Comparisons of Boating and Wading Methods Used to Assess the Status of Flowing Waters. EPA/600/R-00/108. U.S. Environmental Protection Agency, National Exposure Research Laboratory.

Foster, G.R., Meyer, L.D., and Onstad, C.A. (1977). An Erosion Equation Derived from Basic Erosion Principles. *Transactions of the American Society of Agricultural Engineers* 20(4):678–682.

Gibson, G.R., Barbour, M.T., Stribling, J.B., Gerritsen, J., and Karr., J.R. 1996. Biological Criteria: Technical Guidance for Streams and Small Rivers. Washington, DC: U.S. Environmental Protection Agency, Office of Water, EPA 822-B-96-001, 162.

Giddings, E.M.P. 2005. Personal Communication. Email sent to Christine Rohrer. 9 Nov 2005.

Giddings, E.M.P. 2006. Personal Communication. Email sent to Christine Rohrer. 19 June 2006.

Giddings, E.M.P. and Steuer, J.J. 2007. Hydrologic Variability in Urban Streams of the North Carolina Piedmont. In preparation.

Gore, J.A., Layzer, J.B., and Mead, J. 2001. Macroinvertebrate Instream Flow Studies after 20 Years: A Role in Stream Management and Restoration. *Regulated Rivers—Research & Management* 17(4–5):527–542.

Grissinger, E.H. 1982. Bank Erosion of Cohesive Materials. In: Hey, R.D., Bathurst, J.C., and Thorne, C.R. (Editors). *Gravel-Bed Rivers,* p. 273–287. Wiley: Chichester.

Habersack, H.M. 2000. The River-Scaling Concept (RSC): A Basis for Ecological Assessments. *Hydrobiologia* 422/423:49–60.

Hanson, G.J. and Cook, K.R. 2004. Apparatus, Test Procedures, and Analytical Methods to Measure Soil Erodibility *in Situ. Applied Engineering in Agriculture* 20(4):455–462.

Harrelson, C.C., Rawlins, C.L., Potyondy, J.P. 1994. Stream Channel Reference Sites: An Illustrated Guide to Field Technique. *General Technical Reports.* RM-245. Fort Collins, CO: U.S. Department of Agriculture, Forest Service, Rocky Mountain Forest and Range Experiment Station.

Heineman, M. 2007. NETStorm Help. Available from http://www.dynsystem.com/netstorm/ Accessed 7/3/2007.

Henriksen, J.A., Heasley, J., Kennen, J.G., and Nieswand, S. 2006. User's Manual for the Hydoecological Integrity Assessment Process Software (Including the New Jersey Assessment Tools). Fort Collins, CO: U.S. Geological Survey, Fort Collins Science Center. Open-File Report 2006-1093.

Henshaw, P.C. and Booth, D.B. 2000. Natural Restabilization of Stream Channels in Urban Watersheds. *Journal of the American Water Resources Association* 36(6):1219–1236.

Hey, R.D. 1997. Stable River Morphology. In: Thorne, C.T., Hey, R.D., and M.D. Newson (Eds.), *Applied Fluvial Geomorphology for River Engineering and Management*. pp. 223–236 John Wiley & Sons, New York.

Holdsworth, S., Paulsen, S., Gabanski, L. and Barbour, M.T. 2004. Overview of the National Wadeable Streams Assessment Program (WSA). Presented at the 2004 National Monitoring Conference, May 17–20, Chattanooga, TN. National Water Quality Monitoring Council.

Hollis, G.E. 1975. Effect of Urbanization on Floods of Different Recurrence Interval. *Water Resources Research* 11(3):431–435.

Holman-Dodds, J.K., Bradley, A.A., and Potter, K.W. Evaluation of Hydrologic Benefits of Infiltration Based Urban Storm Water Management. 2003. *Journal of the American Water Resources Association*. 39(1):205–215.

Horne, A. and Goldman, C. 1994. *Limnology,* 2nd edition. McGraw-Hill, New York.

Horner, R., May, C.W., Livingston, E., Blaha, D., Scoggins, M., Tims, J., and Maxted, J. 2001. Structural and Non-structural BMPs for Protecting Streams. In: Urbonas, B. (Ed.) *Linking Stormwater BMP Designs and Performance to Receiving Water Impact Mitigation. Engineering Research Foundation Conference, Snowmass, Colorado.* American Society of Civil Engineers, 60–77, Reston, VA.

Huber, W.C. and Dickinson, R.E. 1998. *Storm Water Management Model User's Manual.* Version 4.4, EPA/600/3?88/001a, Environmental Protection Agency, Athens, GA.

Hughes, R.M., Paulsen, S.G. and Stoddard, J.L. 2000. EMAP-Surface Waters: A Multiassemblage Probability Survey of Ecological Integrity in the USA. *Hydrobiologia* 422/423:429–443.

Hynes, H.B.N. 1970. *The Ecology of Running Waters*. University of Toronto Press, Toronto, Ontario, Canada.

Jacobson, R.B., Femmer, S.R. and McKenney, R.A. 2001. *Land-Use Changes and the Physical Habitat of Streams — A Review with Emphasis on Studies within the U.S. Geological Survey Federal-State Cooperative Program.* U.S. Geological Survey, Circular 1175, 63.

Julien, P.Y. 1998. *Erosion and Sedimentation*. Cambridge University Press. New York.

Kappesser, G.B. (2002). A Riffle Stability Index to Evaluate Sediment Loading in Streams. *Journal of the American Water Resources Association* 38(4):1069–1081.

Kauffman, P.R. 1993. Physical Habitat. In: Hughes, R.M. (Editor). *Stream Indicator and Design Workshop.* EPA-600-R-93-138. pp. 59–69Corvallis, OR: U.S. Environmental Protection Agency, Office of Research and Development.

Kaufmann, P.R, Levine, P., Robison, E.G., Seeliger, C., and Peck, D.V. 1999. *Quantifying Physical Habitat in Wadeable Streams.* EPA/620/R-99/003. U.S. Environmental Protection Agency, Washington, D.C.

Kelly, M.H. and Hite, R.L. 1984. *Evaluation of Illinois Stream Sediment Data: 1974–1980.* Illinois Environmental Protection Agency, Division of Water Pollution Control, Springfield, IL.

Kennen, J.G. and Ayers, M.A. 2002. Relation of Environmental Characteristics to the Composition of Aquatic Assemblages Along a Gradient of Urban Land Use in New Jersey, 1996–98. U.S. Geological Survey. *Water-Resources Investigations Report* 02-4069, West Trenton, NJ.

Kerans, B.L. and Karr, J.R. 1994. A Benthic Index of Biotic Integrity (B-IBI) for Rivers of the Tennessee Valley. *Ecological Applications*. 4(4):768–785.

Kirby, C.W. 2003. *Benthic Macroinvertebrate Response to Post-Development Stream Hydrology and Hydraulics.* Doctoral dissertation. George Mason University, Fairfax, VA.UMI Microform 3079343.

Konrad, C.P. and Booth, D.B. 2002. *Hydrologic Trends Resulting from Urban Development in Western Washington Streams*. U.S. Geological Survey Water Resources Investigation Report, 02-4040.

Konrad, C.P., Booth, D.B., and Burges, S.J. 2005. Effects of Urban Development in the Puget Lowland, Washington, on Interannual Streamflow Patterns: Consequences for Channel Form and Streambed Disturbance. *Water Resources Research,* 41.

Konrad, C.P. and Burges, S.J. 2001. Hydrologic Mitigation Using on-Site Residential Storm-Water Detention. *Journal of Water Resources Planning and Management—ASCE* 127(2):99–107.

Konrad, C.P. 2000. Hydrologic Disturbances in Puget Lowland Streams. Doctoral dissertation, University of Washington.

Kwiatkowski, B. 2004. Critical Shear for Cedar Creek. Email communication to N. Postel, August 5, 2004.

Lammert, M. and Allan, J.D. 1999. Environmental Auditing, Assessing Biotic Integrity of Streams: Effects of Scale in Measuring the Influence of Land Use/Cover and Habitat Structure on Fish and Macroinvertebrates. *Environmental Management* 23(2):257–270.

Lamouroux, N., Oliver, J.M., Persat, H., Pouilly, M., Souchon, Y., and Statzner, B. 1999. Predicting Community Characteristics from Habitat Conditions: Fluvial Fish and Hydraulics. *Freshwater Biology* 42:275–299.

Lee, J.G. and Heaney, J.P. 2003. Estimation of Urban Imperviousness and Its Impacts on Storm Water Systems. *Journal of Water Resources Planning and Management — ASCE* 129(5):419–426.

Leopold, L.B. 1968. *Hydrology for Urban Land Planning — A Guidebook on the Hydrologic Effects of Land Use.* U.S. Geological Survey, Circular 554.

MacRae, C.R. 1997. Experience from Morphological Research on Canadian Streams: Is Control of the Two-Year Frequency Runoff Event the Best Basis for Stream Channel Protection? In: Roesner, L.A. (Editor), *Effects of Watershed Development and Management on Aquatic*

Ecosystems, Proceedings of an Engineering Conference, pp. 144–162. American Society of Civil Engineers.

Maddock, I. 1999. The Importance of Physical Habitat Assessment for Evaluating River Health. *Freshwater Biology* 41(2):373–391.

Maddock, I., Thoms, M., Jonson, K., Dyer, F., and Lintermans, M. 2004. Identifying the Influence of Channel Morphology on Physical Habitat Availability for Native Fish: Application to the Two-Spined Blackfish (*Gadopsis bispinosus*) in the Cotter River, Australia. *Marine and Freshwater Research* 55(2):173–184.

Maxted, J. and Shaver, E. 1996. The Use of Retention Basins to Mitigate Stormwater Impacts on Aquatic Life. In: Roesner, L.A. (Editor), *Effects of Watershed Development and Management on Aquatic Ecosystems.* pp. 494–512. Snowbird, UT, American Society of Civil Engineers,

May, C.W. and Horner, R.R. 2002. *The Limitations of Mitigation-Based Stormwater Management in the Pacific Northwest and the Potential of a Conservation Strategy Based on Low-Impact Development Principles. Urban Drainage.* Portland, OR: Sep 8–16. American Society of Civil Engineers.

McCuen, R.H. and Moglen, G.E. 1988. Multicriterion Stormwater Management Methods. *Journal of Water Resources Planning and Management* 144(4):414–431.

McMahon, G., Bales, J.D., Coles, J.F., Giddings, E.M.P., and Zappia, H. 2003. Use of Stage Data to Characterize Hydrologic Conditions in an Urbanizing Environment. *Journal of the American Water Resources Association* 39(6):1529–1546.

McMahon, G. and Cuffney, T.F. 2000. Quantifying Urban Intensity in Drainage Basins for Assessing Stream Ecological Conditions. *Journal of the American Water Resources Association* (JAWRA) 36:1247–1261.

Merigoux, S. and Doledec, S. 2004. Hydraulic Requirements of Stream Communities: A Case Study on Invertebrates. *Freshwater Biology* 49(5):600–613.

Meyer, J.L., Paul, M.J., and Taulbee, W.K . 2005. Stream Ecosystem Function in Urbanizing Landscapes. *Journal of the North American Benthological Society,* 24(3):602–612.

Mid-America Regional Council (MARC) and American Public Works Association (APWA). 2003. *Manual for Best Management Practices for Stormwater Quality.*

Milhous, R.T., Wegner, D.L., Waddle, T. 1984. *User's Guide to the Physical Habitat Simulation System.* FWS/OBS 81/43. U.S. Fish and Wildlife Service, Washington, D.C.

Miltner, R.J., White, D., and Yoder, C.O. 2004. The Biotic Integrity of Streams in Urban and Suburbanizing Landscapes. *Landscape and Urban Planning* 69:87–100.

Moglen, G.E. and R.H. McCuen. 1988. Effects of Detention Basins on In-Stream Sediment Movement. *Journal of Hydrology*. 104: 129–139.

Montgomery, D.R. and Buffington, J.M. 1997. Channel-Reach Morphology in Mountain Drainage Basins. *Geological Society of America Bulletin*. 109: 596–611

Moulton, S.R., II, Kennen, J.G., Goldstein, R.M., and Hambrook, J.A. 2002. Revised Protocols for Sampling Algal, Invertebrate, and Fish Communities as Part of the National Water Quality Assessment Program. U.S. Geological Survey Open-File Report 02-150.

Muhar, S. and Jungwirth, M. 1998. Habitat Integrity of Running Waters: Assessment Criteria and Their Biological Relevance. *Hydrobiologia* 386: 195–202.

Nilsson, C. and Svedmark, M. 2002. Basic Principles and Ecological Consequences of Changing Water Regimes: Riparian Plant Communities. *Environmental Management* 30(4): 468–480.

The Nature Conservancy. 2007. *Indicators of Hydrologic Alteration: User's Manual,* Version 7. Arlington, VA.

North Carolina Department of Environment and Natural Resources (NCDENR). 2006. Standard Operating Procedures for Benthic Macroinvertebrates, Biological Assessment Unit, July 2006. North Carolina Department of Natural Resources, Division of Water Quality, Environmental Sciences Section.

North Carolina Department of Transportation Geographic Information Systems. 2005. GIS Contours and Elevation Data. Accessed 11 March 2005. <http://www.ncdot.org/it/gis/DataDistribution/ContourElevationData/>.

Novotny, V., Braden, J., White, D., Larson, R., Algozin, K., Capodaglio, A., and Schonter, R. 1997. *Use Attainability Analysis: A Comprehensive UAA Technical Reference.* Water Environment Research Foundation Project 91-NPS-1.

Ohio EPA.1987. *Biological Criteria for the Protection of Aquatic Life: Volume II: Users Manual for Biological Field Assessment of Ohio Surface Waters.* OEPA Procedure No. WQMA-SWS-6, November 2, 1987. OEPA, Columbus, OH.

Ohio EPA. 1989. *Biological Criteria for the Protection of Aquatic Life: Volume III: Standardized Biological Field Sampling and Laboratory Methods for Assessing Fish and Macroinvertebrate Communities.* OEPA Procedure No. WQMA-SWS-3, September 30, 1989. OEPA, Columbus, OH.

Ohio EPA. 2003. *Biological and Water Quality Study of the Big Walnut Creek Basin 2000.* EAS/2003-11-10, November 26, 2003.

Ohio EPA. 2005a. *Surface and Ground Water Monitoring and Assessment Strategy 2000–2009, September 20, 2005.* Ohio EPA, Division of Surface Water, Columbus, OH.

Ohio EPA 2005b. *Total Maximum Daily Loads for the Big Walnut Creek Watershed.* Ohio EPA Division of Surface Water, Columbus, OH, August 19, 2005.

Olden, J.D. and Poff, N.L. 2003. Redundancy and the Choice of Hydrologic Indices for Characterizing Streamflow Regimes. *River Research and Applications* 19(2): 101–121.

Osmundson, D.B., Ryel, R.J., Lamarra, V.L., and Pitlick, J. 2002. Flow-Sediment-Biota Relations: Implications for River Regulation Effects on Native Fish Abundance. *Ecological Applications* 12(6): 1719–1739.

Parker, G. In press. Transport of gravel and sediment mixtures, *ASCE Manual 54, Sediment Enginering*, American Society of Civil Engineers, Chapter 3, http://cee.uiuc.edu/people/parkerg/manual_54.htm.

Parsons, M., Thoms, M., and Norris, R. 2002. *Australian River Assessment System: Review of Physical River Assessment Methods — A Biological Perspective, Monitoring River Heath Initiative Technical Report No 21,* Commonwealth of Australia and University of Canberra, Canberra.

Paul, M. J., Stribling, J.B., Klauda, R.J., Kazyak, P.F., and Southerland, M.T. 2002. *A Physical Habitat Index for Freshwater Wadeable Streams in Maryland. Chesapeake Bay and Watershed Programs, Monitoring and Non-Tidal Assessment.* Maryland Department of Natural Resources. CBWP-MANTA-EA-03-4.

Paulson, C., Reeve,. S., Canton, S.P., and Moore, T.F. 2001. *Distinguishing the Relative Influence of Habitat and Water Quality on Aquatic Biota.* Final Report Project 98-WSM-1. Water Environment Research Foundation, Alexandria, VA.

Petts, G.E. 2000. A Perspective on the Abiotic Processes Sustaining the Ecological Integrity of Running Waters. *Hydrobiologia* 422/423: 15–27.

Pfankuch, D.J. 1978. Stream Reach Inventory and Channel Stability Evaluation. United States Department of Agriculture, Forest Service Northern Region.

Plafkin, J.L., Barbour, M.T., Porter, K.D., Gross, S.K., and Hughes, R.M. 1989. *Rapid Bioassessment Protocols for Use in Streams and Rivers: Benthic Macroinvertebrates and Fish.* EPA/444/4-89-001. U.S. Environmental Protection Agency, Office of Water, WA.

Platts, W.S., Megahan, W.F., and Minshall, W.G. 1983. *Methods for Evaluating Stream, Riparian, and Biotic Conditions.* General Technical Report INT-138, USDA Forest Service, Rocky Mountain Research Station, Ogden, UT.

Poff, N.L., Allan, J.D., Bain, M.B., Karr, J.R., Prestegaard, K.L., Richter, B.D., Sparks, R.E., and Stromberg, J.C. 1997. The Natural Flow Regime. *Bioscience* 47(11): 769–784.

Poff, N.L. 1996. A Hydrogeography of Unregulated Streams in the United States and an Examination of Scale-Dependence in Some Hydrological Descriptors. *Freshwater Biology* 36: 71–91.

Poff, N.L. and Ward, J.V. 1989. Implications of Streamflow Variability and Predictability for Lotic Community Structure: A Regional Analysis of Streamflow Patterns. *Canadian Journal of Fisheries and Aquatic Sciences* 46(10): 1805–1818.

Rabeni, C.F. 2000. Evaluating Physical Habitat Integrity in Relation to the Biological Potential of Streams. *Hydrobiologia* 422/423: 245–256.

Rankin, E.T. 1995. *The Use of Habitat Assessments in Water Resource Management Programs, Biological Assessment and Criteria: Tools for Water Resource Planning and Decision Making,* Lewis Publishers, Boca Raton, FL.

Rankin, E.T. 1989. *The Qualitative Habitat Evaluation Index (QHEI): Rationale, Methods, and Application.* Ohio EPA, Columbus, OH.

Reed, T.J., Centinaro, G.L., DeLuca, M.J., Hutchinson, J.T., and Scudder, J. 1997. Water Resources Data for New Jersey—Water Year 1996, Volume 1. Surface Water Data. *U.S. Geological Survey Water Data Report,* NJ-96-1.

Richter, B.D., Baumgartner, J.V., Powell, J., and Braun, D.P. 1996. A Method for Assessing Hydrologic Alteration within Ecosystems. *Conservation Biology* 10: 1163–1174.

Richter, B.D., Baumgartner, J.V., Braun, D.P., and Powell, J. 1998. A Spatial Assessment of Hydrologic Alteration Within a River Network. *Regulated Rivers Research & Management,* 14 (4): 329–340.

Richter, B.D., Baumgartner, J.V., Wigington, R., and Braun, D.P. 1997. How Much Water Does a River Need? *Freshwater Biology* 37(1): 231–249.

Richter, B.D., Mathews, R., and Wigington, R. 2003. Ecologically Sustainable Water Management: Managing River Flows for Ecological Integrity. *Ecological Applications* 13(1): 206–224.

Richter, B.D. and Richter, H.E. 2000. Prescribing Flood Regimes to Sustain Riparian Ecosystems Along Meandering Rivers. *Conservation Biology* 14(5): 1467–1478.

Riis, T. and Biggs, B.J.F. 2003. Hydrologic and Hydraulic Control of Macrophyte Establishment and Performance in Streams. *Limnology and Oceanography* 48(4): 1488–1497.

Roesner, L.A. and Bledsoe, B.P. 2003. *Research Needs: Physical Effects of Wet Weather Flows on Aquatic Habitats.* Water Environment Research Foundation, Alexandria, VA.

Roesner, L.A. and Burgess, E.H. 1991. Hydrology of Urban Runoff Quality Management. Proceedings of the 18th National Conference on Water Resources Planning and Management. Symposium on Urban Water Resources. New Orleans, LA, May 20–22. American Society of Engineers.

Roesner, L.A. (Editor). 1997. Effects of Watershed Development and Management on Aquatic Ecosystems. Proceedings of an Engineering Foundation Conference. Snowbird, UT, August 1996, American Society of Civil Engineers.

Rohrer, C.A. 2004. Modeling the Effect of Stormwater Controls on Sediment Transport in an Urban Stream. Thesis, Department of Civil Engineering, Colorado State University, Fort Collins, CO.

Rohrer, C.A., Postel, N.A., O'Neill, P.A., and Roesner, L.A. 2005. Development of Design Criteria for Regional Stormwater Management Facilities to Maintain Geomorphic Stability in

Cedar Creek. World Water & Environmental Resources Congress: Impacts of Global Climate Change on Water Resources and the Environment. Anchorage, AK, May 15–19.

Rohrer, C.A. and Roesner, L.A. 2005. *Matching the Critical Portion of the Flow Duration Curve to Minimize Changes in Modeled Excess Shear.* 10th International Conference on Urban Drainage, Copenhagen/Denmark, 21–26 August 2005.

Rohrer, C.A. and Roesner, L.A. 2006. Matching the Critical Portion of the Flow Duration Curve to Minimize Changes in Modeled Excess Shear. *Water Science & Technology,* 54:347–354.

Rosgen, D.L. 1994. A Classification of Natural Rivers. *Catena* 22:169–199.

Rosgen, D. 1996. *Applied River Morphology.* Wildlife Hydrology, Pagosa Springs, CO.

Rosgen, D.L. 1999. Development of a River Stability Index for Clean Sediment TMDLs. In: Olsen, D.S. and Potyondy, J.P. (Editors). *Proceedings of Wildland Hydrology*, pp. 25–36. AWRA, Bozeman, MT.

Rosgen, D.L. 2001. *A Practical Method of Computing Streambank Erosion Rate*, Federal Interagency Sediment (FISC) Conference 2001. Reno, NV, March.

Rossman, L.A. 2005. *Storm Water Management Model User's Manual.* Version 5, EPA/600/R-05/040, Environmental Protection Agency, Athens, GA.

Scoggins, M. 2000. *Effects of Hydrologic Variability on Biological Assessments in Streams in Austin, TX.* City of Austin, Watershed Protection Department. NWQMC 2000 proceedings.

Shaver, E. and Ridley, G.. 2002. *Our Use of the Land: Troubling Times and Changing Directions. Urban Drainage.* Portland, OR: Sep 8. American Society of Civil Engineers, 255.

Shields, A. 1936. Anwendung der Aehnlichkeitsmechanik und der Turbulenz Forschung auf die Geschiebebewegung. Berlin: Mitteilungun der Preussische Versuchanstalt für Wasserbau und Schiffbau.

Simon, A. and Hupp, C. R. 1986. Channel Evolution in Modified Tennessee Streams. In: *Proceedings, Fourth Federal Interagency Sedimentation Conference, March 1986,* vol. 2. pp. 71–82. Las Vegas, NV.

Stalnaker, C.B. 1994. Evolution of instream flow habitat modeling, Pages 276–286 in: Calow, P. and Petts, G.E. (Eds.), *The Rivers Handbook: Hydrological and Ecological Principles.* Blackwell Scientific Publications, London.

State Climate Office of North Carolina. 2007. http://www.nc_climate.ncsu.edu/climate/hurrucane.php. Accessed 28 June 2007.

Statzner, B. and Moss, B. 2004. Linking Ecological Function, Biodiversity and Habitat: a Mini-Review Focusing on Older Ecological Literature. *Basic and Applied Ecology* 5(2): 97–106.

Stewardson, M.J. and Gippel, C.J. 2003. Incorporating Flow Variability into Environmental Flow Regimes Using the Flow Events Method. *River Research and Applications* 19(5–6): 459–472.

Stoddard, J.L., Larsen, D.P., Hawkins, C.P., Johnson, R.K., and Norris, R.H. 2006. Setting Expectations for the Ecological Condition of Streams: The Concept of Reference Condition. *Ecological Applications* 16(4): 1267–1276.

Suen, J.P., Herricks, E.E., and Eheart, J.W. 2004. *Ecohydrologic Indicators for Rivers of Northern Taiwan. World Water Congress 2004.* 138, 143. 2004/06/27/. Salt Lake City: American Society of Civil Engineers.

Tate, C.M., Cuffney, T.F., McMahon, G., Giddings, E.M.P., and Zappia, H., 2005. Use of an Urban Intensity Index to Assess Urban Effects on Streams in Three Contrasting Environmental Settings. In: *Effects of Urbanization on Stream Ecosystems,* Brown, L.R., Gray, R.H., Hughes, R.M., and Meador, M.R. (Editors). American Fisheries Society, Bethesda, MD.

Tharme, R.E. 2003. A Global Perspective on Environmental Flow Assessment: Emerging Trends in the Development and Application of Environmental Flow Methodologies for Rivers. *River Research and Applications* 19(5–6): 397–441.

Thomson, J.R., Taylor, M.P., and Brierley, G.J. 2004. Are River Styles Ecologically Meaningful? A Test of the Ecological Significance of a Geomorphic River Characterization Scheme. *Aquatic Conservation–Marine and Freshwater Ecosystems* 14(1): 25–48.

Townsend, C.R., Downes, B.J., Peacock, K., and Arbuckle, C.J. 2004. Scale and the Detection of Land-Use Effects on Morphology, Vegetation and Macroinvertebrate Communities of Grassland Streams. *Freshwater Biology* 49(4): 448–462.

Urban Drainage and Flood Control District (UDFCD). 2001. *Urban Storm Drainage Criteria Manual.* Volumes 1 and 2. Denver.

U.S. Department of Agriculture, Agricultural Research Service, National Sedimentation Laboratory. 2001. *AnnAGNPS Version 2: User Documentation*, Washington D.C.

U.S. EPA 2000. *Stressor Identification Guidance Document.* EPA-822-B-00-025. U.S. Environmental Protection Agency, Washington, D.C.

U.S. EPA. 2002a. *Summary of Biological Assessment Programs and Biocriteria Development for States, Tribes, Territories, and Interstate Commissions: Streams and Wadeable Rivers.* EPA-822-R-02-048. U.S. Environmental Protection Agency.

U.S. EPA. 2002b. *Consolidated Assessment and Listing Methodology: Toward a Compendium of Best Practices, First Edition*. EPA-823-R-03-010. U.S. Environmental Protection Agency, Office of Wetlands, Oceans, and Watersheds.

U.S. EPA. 2003. *Elements of a State Water Monitoring and Assessment Program.* EPA 841-B-03-003. U.S. Environmental Protection Agency, Office of Wetlands, Oceans and Watershed, Assessment and Watershed Protection Division.

U.S. EPA. 2004. *Wadeable Stream Assessment: Field Operations Manual.* EPA 841-B-04-004. U.S. Environmental Protection Agency, Office of Water and Office of Research and Development.

U.S. EPA. 2005. Draft: *Use of Biological Information to Better Define Designated Aquatic Life Uses in State and Tribal Water Quality Standards: Tiered Aquatic Life Uses. Tiered Aquatic Life Uses Document.* EPA-822-R-05-001. U.S. Environmental Protection Agency, Office of Water.

U.S. EPA. 2006. *Wadeable Streams Assessment: A Collaborative Survey of the Nation's Streams.* EPA-841-B-06-002. U.S. Environmental Protection Agency, Office of Water.

USGS. 2003. National Land Cover Database Zone 60 Imperviousness Layer. Edition 1.0. U.S. Geological Survey. Sioux Falls, SD.

USGS. 2006a. National Elevation Dataset (NED) 1/3 Arc Second (~10m) Digital Elevation Map. U.S. Geological Survey. Sioux Falls, SD.

USGS. 2006b. Raleigh Durham (Mar 2002) Orthoimagry. U.S. Geological Survey. Sioux Falls, SD.

USGS. 2007. National Water Quality Assessment Program — Effects of Urbanization on Stream Ecosystems. http://co.water.usgs.gov/nawqa/EUSE/ Web site accessed 2/20/07.

Wake County North Carolina. Tax/Parcel, Digital Orthophotos, Topographic & Soils Maps. 1999. Accessed 11 March 2005 from http://lnweb02.co.wake.nc.us/gis/gismaps.nsf.

Walsh, C.J. 2004. Protection of in-Stream Biota from Urban Impacts: Minimise Catchment Imperviousness or Improve Drainage Design? *Marine and Freshwater Research* 55(3): 317–326.

Walsh, C.J., Papas, P.J., Crowther, D., Sim, P.T., and Yoo, J. 2004. Stormwater Drainage Pipes as a Threat to a Stream-Dwelling Amphipod of Conservation Significance, *Austrogammarus australis*, in South-Eastern Australia. *Biodiversity and Conservation,* 13: 781–793.

Walsh, C.J., Roy, A.H., Feminella, J.W., Cottingham, P.D., Groffman, P.M. and R.P. Morgan II. 2005. The Urban Stream Syndrome: Current Knowledge and the Search for a Cure. *Journal of the North American Benthological Society,* 24(3): 706–723.

Ward, J.V., Malard, F., and Tockner, K. 2002. Landscape Ecology: A Framework for Integrating Pattern and Process in River Corridors. *Landscape Ecology* 17(Suppl.1): 34–45.

Water Environment Federation (WEF) and American Society of Civil Engineers. 1998. *Urban Runoff Quality Management. WEF Manual of Practice;* no. 23. *ASCE Manual and Report on Engineering Practice* 87. Alexandria, VA.

Waters, T.F. 1995. Sediment in Streams: Sources, Biological Effects and Control. *American Fisheries Society Monograph 7.* Bethesda, MD: American Fisheries Society.

Winward, A.H. 2000. *Monitoring the Vegetation Resources in Riparian Areas.* General Technical Report RMRSGTR-47. Ogden, UT: U.S. Department of Agriculture, Forest Service, Rocky Mountain Research Station.

Wolman, M.G. 1954. A Method for Sampling Coarse River-Bed Material. *Transactions of the American Geophysical Journal* 35(6): 951–956.

Wu, F.C. and Wang, C.F. 2002. Effect of Flow-Related Substrate Alteration on Physical Habitat: A Case Study of the Endemic River Loach *Sinogastromyzon puliensis* (Cypriniformes, Homalopteridae) Downstream of Chi-Chi Diversion Weir, Chou-Shui Creek, Taiwan. *River Research and Applications* 18(2): 155–169.

Yoder, C.O. 1998. Important Concepts and Elements of an Adequate State Watershed Monitoring and Assessment Program. In: *Proceedings of the NWQMC National Conference on Monitoring: Critical Foundations to Protecting Our Waters,* pp. 615–628. U.S. Environmental Protection Agency, Washington, D.C.

Yoder, C.O. 2003. *Region V State Bioassessment and Ambient Monitoring Programs: Initial Evaluation and Review.* Technical Report MBI/01-03-1. Midwest Biodiversity Institute, Columbus, OH.

Yoder, C.O. and Rankin, E.T. 1995a. Biological Criteria Program Development and Implementation in Ohio. In: W.S. Davis and T.P. Simon (editors). *Biological Assessment and Criteria: Tools for Water Resource Planning and Decision Making*. pp. 109–144. Lewis Publishers, Boca Raton, FL.

Yoder, C.O. and Rankin, E.T. 1995b. Biological Response Signatures and the Area of Degradation Value: New Tools for Interpreting Multimetric Data. In: Davis, W.S., and Simon, T.P. (editors). *Biological Assessment and Criteria: Tools for Water Resource Planning and Decision Making*. pp. 263–286. Lewis Publishers, Boca Raton, FL.

Yoder, C.O., Miltner, R., and White, D. 2000. Using Biological Criteria to Assess and Classify Urban Streams and Develop Improved Landscape Indicators. In: *Proceedings of the National Conference on Tools for Urban Water Resource Management & Protection, Chicago, IL, Conference Draft,* pp. 56–82.. EPA/625/R-00/001.

WERF SUBSCRIBERS

WASTEWATER UTILITY

Alabama
Montgomery Water Works & Sanitary Sewer Board

Alaska
Anchorage Water & Wastewater Utility

Arizona
Glendale, City of, Utilities Department
Mesa, City of
Peoria, City of
Phoenix Water Services Dept.
Pima County Wastewater Management
Safford, City of

Arkansas
Little Rock Wastewater Utility

California
Central Contra Costa Sanitary District
Corona, City of
Crestline Sanitation District
Delta Diablo Sanitation District
Dublin San Ramon Services District
East Bay Dischargers Authority
East Bay Municipal Utility District
Eastern Municipal Water District
El Dorado Irrigation District
Fairfield-Suisun Sewer District
Fresno Department of Public Utilities
Inland Empire Utilities Agency
Irvine Ranch Water District
Las Virgenes Municipal Water District
Livermore, City of
Los Angeles, City of
Los Angeles County, Sanitation Districts of
Napa Sanitation District
Orange County Sanitation District
Palo Alto, City of
Riverside, City of
Sacramento Regional County Sanitation District
San Diego Metropolitan Wastewater Department, City of
San Francisco, City & County of
San Jose, City of
Santa Barbara, City of
Santa Cruz, City of
Santa Rosa, City of
South Bayside System Authority
South Coast Water District
South Orange County Wastewater Authority
Stege Sanitary District
Sunnyvale, City of
Union Sanitary District
West Valley Sanitation District

Colorado
Aurora, City of
Boulder, City of
Greeley, City of
Littleton/Englewood Water Pollution Control Plant
Metro Wastewater Reclamation District, Denver

Connecticut
Greater New Haven WPCA
Stamford, City of

District of Columbia
District of Columbia Water & Sewer Authority

Florida
Broward, County of
Fort Lauderdale, City of
Jacksonville Electric Authority (JEA)
Miami-Dade Water & Sewer Authority
Orange County Utilities Department
Reedy Creek Improvement District
Seminole County Environmental Services
St. Petersburg, City of
Tallahassee, City of
Tampa, City of
Toho Water Authority
West Palm Beach, City of

Georgia
Atlanta Department of Watershed Management
Augusta, City of
Clayton County Water Authority
Cobb County Water System
Columbus Water Works
Fulton County
Gwinnett County Department of Public Utilities
Savannah, City of

Hawaii
Honolulu, City & County of

Idaho
Boise, City of

Illinois
Greater Peoria Sanitary District
Kankakee River Metropolitan Agency
Metropolitan Water Reclamation District of Greater Chicago
Wheaton Sanitary District

Iowa
Ames, City of
Cedar Rapids Wastewater Facility
Des Moines, City of
Iowa City

Kansas
Johnson County Wastewater
Unified Government of Wyandotte County/ Kansas City, City of

Kentucky
Louisville & Jefferson County Metropolitan Sewer District
Sanitation District No. 1

Louisiana
Sewerage & Water Board of New Orleans

Maine
Bangor, City of
Portland Water District

Maryland
Anne Arundel County Bureau of Utility Operations
Howard County Bureau of Utilities
Washington Suburban Sanitary Commission

Massachusetts
Boston Water & Sewer Commission
Massachusetts Water Resources Authority (MWRA)
Upper Blackstone Water Pollution Abatement District

Michigan
Ann Arbor, City of
Detroit, City of
Holland Board of Public Works
Saginaw, City of
Wayne County Department of Environment
Wyoming, City of

Minnesota
Rochester, City of
Western Lake Superior Sanitary District

Missouri
Independence, City of
Kansas City Missouri Water Services Department
Little Blue Valley Sewer District
Metropolitan St. Louis Sewer District

Nebraska
Lincoln Public Works and Utilities Department

Nevada
Henderson, City of
Las Vegas, City of
Reno, City of

New Jersey
Bergen County Utilities Authority
Ocean County Utilities Authority
Passaic Valley Sewerage Commissioners

New York
New York City Department of Environmental Protection

North Carolina
Charlotte/Mecklenburg Utilities
Durham, City of
Metropolitan Sewerage District of Buncombe County
Orange Water & Sewer Authority
University of North Carolina, Chapel Hill

Ohio
Akron, City of
Butler County Department of Environmental Services
Columbus, City of
Metropolitan Sewer District of Greater Cincinnati
Northeast Ohio Regional Sewer District
Summit, County of

Oklahoma
Oklahoma City Water & Wastewater Utility Department
Tulsa, City of

Oregon
Albany, City of
Clean Water Services
Eugene, City of
Gresham, City of
Portland, City of Bureau of Environmental Services
Water Environment Services

Pennsylvania
Hemlock Municipal Sewer Cooperative (HMSC)
Philadelphia, City of
University Area Joint Authority

South Carolina
Charleston Water System
Mount Pleasant Waterworks & Sewer Commission
Spartanburg Water

Tennessee
Cleveland Utilities
Knoxville Utilities Board
Murfreesboro Water & Sewer Department
Nashville Metro Water Services

Texas
Austin, City of
Dallas Water Utilities
Denton, City of
El Paso Water Utilities
Fort Worth, City of
Houston, City of
San Antonio Water System
Trinity River Authority

Utah
Salt Lake City Corporation

Virginia
Alexandria Sanitation Authority
Arlington, County of
Fairfax County
Hampton Roads Sanitation District
Hanover, County of
Henrico, County of
Hopewell Regional Wastewater Treatment Facility
Loudoun Water
Lynchburg Regional WWTP
Prince William County Service Authority
Richmond, City of
Rivanna Water & Sewer Authority

Washington
Everett, City of
King County Department of Natural Resources
Seattle Public Utilities
Sunnyside, Port of
Yakima, City of

Wisconsin
Green Bay Metro Sewerage District
Kenosha Water Utility
Madison Metropolitan Sewerage District
Milwaukee Metropolitan Sewerage District
Racine, City of
Sheboygan Regional Wastewater Treatment
Wausau Water Works

Australia
ACTEW (Ecowise)
South Australian Water Corporation
South East Water Limited
Sydney Water Corporation
Water Corporation of Western Australia

Canada
Edmonton, City of/Edmonton Waste Management Centre of Excellence
Lethbridge, City of
Regina, City of, Saskatchewan
Toronto, City of, Ontario
Winnipeg, City of, Manitoba

New Zealand
Watercare Services Limited

STORMWATER UTILITY

California
Fresno Metropolitan Flood Control District
Los Angeles, City of, Department of Public Works
Monterey, City of
San Francisco, City & County of
Santa Rosa, City of
Sunnyvale, City of

Colorado
Aurora, City of
Boulder, City of

Florida
Orlando, City of

Georgia
Griffin, City of

Iowa
Cedar Rapids Wastewater Facility
Des Moines, City of

Kansas
Overland Park, City of

Kentucky
Louisville & Jefferson County Metropolitan Sewer District

Maine
Portland Water District

North Carolina
Charlotte, City of, Stormwater Services

Pennsylvania
Philadelphia, City of

Tennessee
Chattanooga Stormwater Management

Texas
Harris County Flood Control District, Texas

Washington
Bellevue Utilities Department
Seattle Public Utilities

STATE

Arkansas Department of Environmental Quality
Connecticut Department of Environmental Protection
Kansas Department of Health & Environment
Kentucky Department of Environmental Protection
New England Interstate Water Pollution Control Commission (NEIWPCC)
Ohio River Valley Sanitation Commission
Urban Drainage & Flood Control District, CO

CORPORATE

ADS Environmental Services
Advanced Data Mining International
Alan Plummer & Associates
Alpine Technology Inc.
Aqua-Aerobic Systems Inc.
Aquateam–Norwegian Water Technology Centre A/S
ARCADIS
Associated Engineering
Black & Veatch
Blue Water Technologies, Inc.
Boyle Engineering Corporation
Brown & Caldwell
Burgess & Niple, Ltd.
Burns & McDonnell
CABE Associates Inc.
The Cadmus Group
Camp Dresser & McKee Inc.
Carollo Engineers Inc.
Carpenter Environmental Associates Inc.
Carter & Burgess, Inc.
CET Engineering Services
CH2M HILL
CRA Infrastructure & Engineering
CONTECH Stormwater Solutions
D&B/Guarino Engineers, LLC
Damon S. Williams Associates, LLC
Earth Tech Inc.
Ecovation
EMA Inc.
Environmental Operating Solutions, Inc.
Environ/The ADVENT Group, Inc.
Fay, Spofford, & Thorndike Inc.
Freese & Nichols, Inc.
ftn Associates Inc.
Gannett Fleming Inc.
Garden & Associates, Ltd.
Geosyntec Consultants
GHD
Golder Associates Ltd.
Greeley and Hansen LLC
Hazen & Sawyer, P.C.
HDR Engineering Inc.
HNTB Corporation
Hydromantis Inc.
HydroQual Inc.
Infilco Degremont Inc.
Jacques Whitford NAWE, Inc.
Jason Consultants LLC Inc.
Jordan, Jones, & Goulding Inc.
KCI Technologies Inc.
Kelly & Weaver, P.C.
Kennedy/Jenks Consultants
KMK Consultants
Komline Sanderson Engineering Corporation
Larry Walker Associates
Limno-Tech Inc.
The Low Impact Development Center Inc.
Malcolm Pirnie Inc.
Material Matters
McKim & Creed
Metcalf & Eddy Inc.
Monteco Corporation
MPR Engineering Corporation, Inc.
MWH
O'Brien & Gere Engineers Inc.
Odor & Corrosion Technology Consultants Inc.
Original Engineering Consultants, Ltd.
Paradigm Environmental Technologies, Inc.
Parametrix Inc.
Parsons
Post, Buckley, Schuh & Jernigan
Praxair, Inc.
Ring Industrial Group
RMC Water & Environment
Ross & Associates Ltd.
SAIC
Siemens Water Technologies
The Soap & Detergent Association
Smith & Loveless, Inc.
Stantec Consulting Inc.
Stearns & Wheler, LLC
Stone Environmental Inc.
Stratus Consulting Inc.
Synagro Technologies Inc.
Tetra Tech Inc.
Trojan Technologies Inc.
Trussell Technologies, Inc.
URS Corporation
Westin Engineering Inc.
Weston Solutions Inc.
Woodard & Curran
Zoeller Pump Company

INDUSTRY

American Electric Power
American Water
Anglian Water Services, Ltd.
ChevronTexaco Energy Research & Technology Company
The Coca-Cola Company
Dow Chemical Company
DuPont Company
Eastman Chemical Company
Eli Lilly & Company
Merck & Company Inc.
Procter & Gamble Company
Suez Environnment
United Utilities North West (UUNW)
United Water Services LLC

Note: List as of 4/10/08

WERF Board of Directors

WERF Research Council

IWA Publishing's authorised EU representative for General Product Safety Regulations is Diane D'Arras, 15 rue Duret, 75116 Paris, France, e-mail: safety@iwap.co.uk.

Printed and bound by CPI Group (UK) Ltd, Croydon, CR0 4YY
16/06/2026
02140383-0001